# SpringerBriefs in Philosophy

SpringerBriefs present concise summaries of cutting-edge research and practical applications across a wide spectrum of fields. Featuring compact volumes of 50 to 125 pages, the series covers a range of content from professional to academic. Typical topics might include:

- A timely report of state-of-the art analytical techniques
- A bridge between new research results, as published in journal articles, and a contextual literature review
- A snapshot of a hot or emerging topic
- An in-depth case study or clinical example
- A presentation of core concepts that students must understand in order to make independent contributions

SpringerBriefs in Philosophy cover a broad range of philosophical fields including: Philosophy of Science, Logic, Non-Western Thinking and Western Philosophy. We also consider biographies, full or partial, of key thinkers and pioneers.

SpringerBriefs are characterized by fast, global electronic dissemination, standard publishing contracts, standardized manuscript preparation and formatting guidelines, and expedited production schedules. Both solicited and unsolicited manuscripts are considered for publication in the SpringerBriefs in Philosophy series. Potential authors are warmly invited to complete and submit the Briefs Author Proposal form. All projects will be submitted to editorial review by external advisors.

SpringerBriefs are characterized by expedited production schedules with the aim for publication 8 to 12 weeks after acceptance and fast, global electronic dissemination through our online platform SpringerLink. The standard concise author contracts guarantee that

- an individual ISBN is assigned to each manuscript
- each manuscript is copyrighted in the name of the author
- the author retains the right to post the pre-publication version on his/her website or that of his/her institution.

More information about this series at https://link.springer.com/bookseries/10082

Robert Greenleaf Brice

# Wittgenstein's On Certainty: Insight and Method

Springer

Robert Greenleaf Brice
Department of Sociology, Anthropology, & Philosophy
Northern Kentucky University
Highland Heights, KY, USA

ISSN 2211-4548 ISSN 2211-4556 (electronic)
SpringerBriefs in Philosophy
ISBN 978-3-030-90783-9 ISBN 978-3-030-90781-5 (eBook)
https://doi.org/10.1007/978-3-030-90781-5



This Springer imprint is published by the registered company Springer Nature Switzerland AG
The registered company address is: Gewerbestrasse 11, 6330 Cham, Switzerland

*For Pat Bourgeois, a wonderful mentor, colleague, and friend.*

# Preface

In *On Certainty*, the important, but to many readers obscure, twentieth century Austrian philosopher, Ludwig Wittgenstein, provides not only a brilliant solution to a previously intractable philosophical problem but also the elements of an entirely new way of approaching this and similar longstanding, apparently unresolvable, problems. In these notes, he reconceives the problem of *radical skepticism*—the claim that we can never really be certain of anything except the contents of our own minds—as a kind of philosophical "disease" of thought. His approach to the problem, which I will emphasize is similar to the treatment of disease, has two main goals:

1. bring about an awareness in the philosopher that this kind of extreme skepticism is not a methodological approach to be taken seriously, and, with this awareness,
2. an attempt to replace this radical skepticism with a practical, Common Sense framework.

Implicit in Wittgenstein's approach are a number of strategies found in a contemporary approach to psychotherapy known as *Cognitive Behavioral Therapy (CBT)*. These strategies, along with philosophical methods and scientific practices rooted in the *Scottish School of Common Sense*, seek to diagnose and treat irrational thoughts and beliefs that often emerge (and re-emerge) in the discipline of philosophy. The aim is to provide the philosopher with tools necessary to adjust and reshape these irrational, self-defeating thoughts and beliefs into something new, something healthy.

CBT did not exist (in name) in Wittgenstein's day; however, the parallels between strategies found in the theory and Wittgenstein's treatment of radical skepticism are strikingly similar. I intend to draw on these similarities in examining and explaining *On Certainty*. This examination will take focus on:

- The historical background against which *On Certainty* was written
- The topics Wittgenstein hoped to resolve/dissolve by applying his therapeutic approach
- An explanation of important individual parts, sections, and statements

The book will serve first and foremost as a guide to the reader's understanding of Wittgenstein's often opaque, yet brilliant and philosophically liberating, dissolution of the problem of radical skepticism. Secondarily, it will illuminate Wittgenstein's conception of philosophy as something like psychological therapy.

The book is divided into seven chapters. Chapter 1 offers a brief description of *On Certainty*'s development, the secondary literature surrounding it, its placement in the canon, and a short section on the philosopher's succumbing to a certain kind of "philosophical disease." Chapter 2 provides a brief introduction to the notion of philosophical therapy, its connection to eighteenth-century *Scottish School of Common Sense*, and further back still to its connection with ancient Greek philosophy. Chapter 3 focuses on three related issues: (a) the extreme form of skepticism developed by Rene Descartes, (b) G.E. Moore's Common Sense response to it, and (c) Wittgenstein's reply to both. Chapter 4 offers an account of Wittgenstein's general therapeutic approach as it applies to knowledge and the skeptical challenge before turning to its application in *On Certainty*. In Chap. 5, I consider Wittgenstein's rejection of the conventional understanding of the relationship between knowledge and certainty, and assess his reshaping of our traditional conception of belief. In Chap. 6, I turn to the relationship between the "language game" of knowledge, "hinge-propositions," and "actional certainty" (terms to be explained), as well as the relationship between two different ways we acquire certainty which I characterize as "Bottom-Up" and "Top-Down." Finally, in Chap. 7, I turn to philosophical therapy as a method, an account of what, generally, Wittgenstein's method can do for us, and what needs to be done in order for this to happen.

Highland Heights, KY, USA Robert Greenleaf Brice

*A philosopher is a man who has to cure many intellectual diseases in himself before he can arrive at the notions of common sense.*

Ludwig Wittgenstein

# Acknowledgments

In fall 1999, I was a graduate student at Michigan State University, enrolled in Martin Benjamin's seminar on Ludwig Wittgenstein. For many of us in the class, it was the first time we had read *On Certainty*. I was fascinated. Benjamin was quite familiar with the secondary literature and adroitly helped us make sense of it all. Fortunately for me, he later agreed to supervise my dissertation. This served as the germ for my first article, "Recognizing Targets." Both Benjamin's teaching and writing have had an enduring influence on my work, including this manuscript. Indeed, if not for Martin and his insightful questions and astute suggestions, this book would not have been completed.

I would also like to extend a special thanks to my friend and former professor, Paul Streveler. Paul provided me with invaluable comments, important criticisms, and penetrating analysis throughout the course of this multi-year project.

In addition to Martin and Paul, I have discussed the topics found here with a variety of colleagues, students, friends, and family members. I have learned a lot from all of them. In particular, I would like to thank Patrick Bourgeois, Augustine Yaw Frimpong-Mansoh, Jonathan Reynolds, Michael Bush, Nigel Pleasants, William Hirstein, Rudy Garns, Alison Reiheld, Akinori Hayashi, Patrick Edwards, Isabel Sleczkowski, Blake Jaggars, Russ Gifford, Cruz Carrillo, Aaron Devers, Madison West, Zachary Long, and my mom, Suzanne Brice. I apologize to anyone I have forgotten to include here.

I would also like to thank my editor at Springer Nature, Deepthi Vasudevan, as well as editorial assistant, Svetlana Kleiner, for not only shepherding my manuscript through the various stages but for providing me the extra time needed to complete it. After 6 years, including two substantial rewrites and one complete overhaul, I have produced a relatively short book. However, lots of work went into making it this way.

I would also like to thank the editors and publishers who gave permission to reprint selections found in the book. In particular, I would like to thank *Lexington Books* for allowing me to reuse selections from Chaps. 1, 3, and 9 of my book, *Exploring Certainty*. I would also like to thank *Wiley Publishing* for permitting me to include selections from Wittgenstein's *On Certainty*.

Finally, my deepest gratitude is reserved for my wife, Lori Hazen, and for my daughter Cecelia Brice. They have celebrated my highs and endured my lows throughout this project. I am quite fortunate to have such wonderful people in my life.

# Contents

# Chapter 1
# *On Certainty*: Scholarship, Development, and Placement

**Abstract** This chapter offers a brief description of *On Certainty*'s development, the secondary literature surrounding it, and its placement in the canon. It also includes a short section on the philosopher's succumbing to a certain kind of "philosophical disease."

**Keywords** Certainty · Knowledge · Scholarship · Canon · Therapeutic · G.E. Moore · Ludwig Wittgenstein

## 1.1 Can We Know Anything with Certainty?

Can we know anything *with certainty*? This is a fitting question given that certainty, or perceived certainty, is involved in much of what we do. Possessing it implies that we not only *know* something, but that we know it with the *highest level of confidence*. Certainty then appears to represent a degree of knowledge—the highest degree.

This is a central topic of interest among philosophers. Throughout the centuries, they have expressed this level of confidence with some variation of the phrase, "I know *with certainty* that *p*," where "p" represents some statement or proposition such as "2 + 2 = 4" or "grass is green." For example,

- In his *Summa Theologiae*, St. Thomas Aquinas described his conviction that God exists as a "certitude *of knowledge*" (Aquinas 1981: 3).
- In *Meditations on First Philosophy*, Rene Descartes promised to "put aside everything that admits of the least of doubt…until [he] *knew something certain*" (Descartes 1993: AT VII 24).
- In *An Essay Concerning Human Understanding*, John Locke says that men can "attain to all the knowledge they have, without the help of any innate impressions; and may *arrive at certainty*, without such original notions or principles (Locke 1975: 48).



R. G. Brice, *Wittgenstein's On Certainty: Insight and Method*, SpringerBriefs in Philosophy, https://doi.org/10.1007/978-3-030-90781-5_1

- In his book *Generales Inquisitiones*, while referring to God's knowledge of contingent truths, Gottfried Leibniz says, God "knows [these] truths *with certainty*" (Leibniz 1982: 77).[1]
- In his *Enquiry Concerning Human Understanding*, while discussing human conduct and necessity, David Hume says, "*I know with certainty* that my friend will not put his hand into the fire and hold it there until it is consumed" (Hume 2007: 67).[2]
- In *Problems of Philosophy*, Bertrand Russell asks whether there is "any *knowledge* in the world *so certain* that no reasonable man could doubt it?" (Russell 1912: 1).
- In "Defense of Common Sense," G.E. Moore begins his article "by enunciating…a whole long list of propositions…every one of which," he says, "I know, *with certainty*, to be true" (Moore 1925: 32).

Other philosophers have offered similar declarations.[3] It seems clear certainty has, for many centuries, represented a degree—the *highest* degree—of knowledge.

In a series of exploratory notes, written at the end of his life, Ludwig Wittgenstein challenged this long-standing and widely accepted view. Published posthumously as *On Certainty*, these notes regard certainty not as a degree of knowledge but *distinct* from it; knowledge and certainty, he tells us, belong to "different categories" (Wittgenstein 1972: §308). While knowledge is both propositional and ratiocinative, true and demonstrably justifiable, certainty possesses none of these characteristics. Fundamental certainty, according to Wittgenstein, is borne out not in what we say, think, or write, but rather in what we unreflectively *do*; exhibited in our instinctual and habitual actions and deeds. He draws out this distinction in many passages in *On Certainty*, for instance,

> §148: Why do I not satisfy myself that I have two feet when I want to get up from a chair? There is no why. I simply don't. This is how I *act*.

[1] Also see Leibniz, *New Essays On Human Understanding*, (1703–5), (trans. and eds.) Remnant & Bennett, A VI, 6. Book 4—On Knowledge: 444–5: "Certainty might be taken to be knowledge of a truth such that to doubt it in a practical way would be insane … " Scholars, like Julia Weckend, have argued that Leibniz's use of "moral certainty" stands for "the highest degrees of knowledge obtainable by humans in a particular domain at a time." "Leibniz on Certainty," *Tercentenary Essays on the Philosophy and Science of Leibniz* pp. 203–227.

[2] In *Philosophical Investigations* §474, Wittgenstein, seemingly in conversation with Hume here, says: "I shall get burnt if I put my hand in the fire: that is certainty. That is to say: here we see the meaning of certainty. (What it amounts to, not just the meaning of the word 'certainty.')"

[3] For instance, Edmund Husserl, while criticizing the Cartesian mediator, said, "they must carry with them an absolute certainty…governed by the idea of a definitive system of knowledge." Husserl, Edmund, *Cartesian Meditations*, translated by Dorion Cairns, The Hague Nijhoff, 1969. Reprinted, Certainty, Hackett Publishing. p 51. "Logical analysis," according to Hans Reichenbach, "may be employed for the establishment of a kind of knowledge which is regarded as absolutely certain…" *The Rise of Scientific Philosophy* (1973). Berkeley: University of California Press. p.34. In *City of God*, St. Augustine wrote, "I am most certain that I am, and that I know and delight in this." *City of God*, St. Augustine, Marcus Dods (trans.), Book XI, Chapter 26, DigiReads Publishing.com, p. 316.

> §359: I want to conceive it [certainty] as something that lies beyond being justified or unjustified; as it were, as something animal.
> §414: What we have here [certainty] is a foundation for all of my action. But it seems to me that it is wrongly expressed by the words 'I know.'

This *actional* certainty not only underpins knowledge, it "forms the foundation of *all* [of our] operating with thoughts (with language)" (Wittgenstein 1972: §401). *On Certainty* not only alters our understanding of certainty, this first draft set of notes identifies and presents us with an original kind of foundation for knowledge.

Over the past half-century since its publication, the scope and breadth of scholarship on these notes has narrowed. This is due, in large part, to a renewed effort to read *On Certainty* less as an extension of thoughts and ideas found in his earlier work, *Philosophical Investigations*, and more as marking a new period of thought. The remainder of this chapter will focus on issues of scholarship surrounding *On Certainty*.

## 1.2 The 3 Phases of *On Certainty* Scholarship

2019 marked the 50th anniversary of the publication of Ludwig Wittgenstein's *On Certainty*. After 50 years, one might expect to find a sizeable number of studies devoted to this important work, but for roughly the first 25 of those years—from the early 1970s to the mid-1990s—there were only seven books published. Avrum Stroll divides this period into two phases:

> There were five books in what might be called an extended first phase…those published between 1971 and 1981… This brief, concentrated flurry stopped suddenly, and no books on the Wittgenstein opus appeared until 1989 (Stroll 1994: 7–8).[4]

Phase two (1989), produced the remaining two books.[5] This is a surprisingly low number given how long the work had been in circulation.[6] However, thanks to Stroll's own book, *Moore and Wittgenstein on Certainty* (1994), scholarship into *On Certainty* entered a third phase.[7]

---

[4] This "extended first phase" includes: Malgren, Helge. 1971. *Intentionality and Knowledge: Studies in the Philosophy of G.E. Moore and Ludwig Wittgenstein*. Goteborg: Goteborgs Universitet; Wilde, Carolyn. 1976. *Certainty: A Discussion of Wittgenstein's Notes in On Certainty*. London: Open University Press; Wolgast, Elizabeth. 1977. *Paradoxes of Knowledge*. Ithaca: Cornell University Press; Svensson, Gunnar. 1981. *On Doubting the Reality of Reality: Moore and Wittgenstein on Sceptical Doubts*. Stockholm: Almqvist & Wiksell International.

[5] McGinn, Marie. 1989. *Sense and Certainty*. Oxford: Oxford University Press. Conway, Gertrude. 1989. *Wittgenstein On Foundations* Atlantic Highlands: Humanities Press International.

[6] By way of comparison, in the 25 years from 1953 until 1978, the number of books devoted to Wittgenstein's *Philosophical Investigations* was 44. See Frongia, G. and McGuiness, B. 1990. *Wittgenstein: A Bibliographical Guide*. Oxford: Basil Blackwell Press.

[7] In addition to Stroll, this third group includes Daniele Moyal-Sharrock, Analisa Coliva, and Robert Greenleaf Brice, among others.

In the first two phases, scholars either evaluated *On Certainty* in light of Wittgenstein's other work, *Philosophical Investigations*, or they failed to recognize *On Certainty*'s originality. In the third phase, however (1994-present), closer attention was paid to the way the work progressed. Researchers noticed Wittgenstein's move away from thinking of certainty strictly in propositional terms and towards thinking of it as something different, something *actional*; that is, grounded in human action. Propositions of the form, "I know that *p*," belong to what Wittgenstein calls "the language-game of knowledge," but certainty is something different—often playing a foundational role, grounding this and our many other *language-games*. (Readers unfamiliar with this concept, central to an understanding of Wittgenstein, are advised to turn to the Appendix, where the notion is more fully explained. There is also a brief explication below, in Sect. 1.4.) Stroll, in particular, was the first to see the importance of Wittgenstein's emphasizing a categorical difference between knowledge and certainty. He was also the first to suggest that what Wittgenstein was arguing for was "a highly original form of foundationalism."

> It is Wittgenstein's main thesis in *On Certainty* that what stands fast is not subject to justification, proof, the adducing of evidence, or doubt, and is neither true nor false. Whatever is subject to these ascriptions belongs to the language-game. But certitude is not so subject, and therefore it stands outside of the language-game (Stroll 1994: 138).

What "stands fast" for us are not propositions, but deeper sorts of convictions—convictions that are non-linguistic—reflexive—and expressed in action. First and foremost, these convictions are located in what we *do*, not in what we say.

Daniele Moyal-Sharrock's work here is also significant. Apart from Stroll, Moyal-Sharrock is perhaps the one scholar who has done the most to highlight the significance of *On Certainty*. She also deserves recognition for the classification of what is now referred to as the *Third Wittgenstein*, discussed below.

While the quality of scholarship in the first two phases was unremarkable, the question remains: why were there so few books written?[8] Perhaps, as Stroll charges,

> many scholars did not fully grasp the importance of *On Certainty* because it could be understood only by reading it through the filter of [G.E.] Moore's essays, written more than a quarter of a century earlier, and this they had no interest in doing (Stroll 1994: 8).

[8] There are a number of reasonable explanations to account for the lack of attention paid to *On Certainty*, chief among them is the fact that, given the enormous amount of material left to his literary executors, the work lay buried in the huge Wittgensteinian *nachlass* until the late 1960s. And, as was already mentioned, there was an incredible amount of attention paid to *Philosophical Investigations*, published just two years after his death (1953). As the *Investigations* had been revised and polished by Wittgenstein, suggesting that it was almost ready for publication, it held a kind of importance that eclipsed everything else he wrote.

This seems reasonable.[9] *On Certainty* is, after all, a reaction to G.E. Moore's articles, "A Defense of Common Sense" and "Proof of an External World."[10] Without a thorough understanding of Moore, his Common Sense approach, and what (whom) he is responding to, *On Certainty* makes for a difficult set of notes to decipher. Reluctance among scholars to do background research, the sort necessary to give *On Certainty* its proper context, might help explain why the number of studies devoted to this work during phases I and II are so few.

## 1.3 How *On Certainty* Came to Be Written

In 1949, two years before his death, Wittgenstein accepted an invitation from his former student, Norman Malcolm, to visit him at Cornell University in Ithaca, New York. While in America, the two had several conversations about Moore's use of knowledge claims, specifically of the form: "I *know* that *p*" and "I *know* with *certainty* that *p*." As Malcolm says in his memoir of Wittgenstein:

> The discussions that were of most value to me that summer were a series that took place between Wittgenstein and me, our topic being Moore's...insistence that it is a correct use of language for him to say, when holding one of his hands before him, "I *know* that this is a hand;" or to say, while pointing at a tree a few feet away, "I *know* for *certain* that this is a *tree*!"...Wittgenstein and I discussed these matters in a number of conversations, he making many observations of the first importance about the concept of knowledge (Malcolm 1962: 87).

After returning to Europe, Wittgenstein began recording his thoughts in notebooks, later published posthumously as *Über Gewissheit* (*On Certainty*).

The work is divided into four parts, written at different times and in different locations. In the *preface*, G.E.M. Anscombe recalls that Part I was probably written in Vienna:

> the first part [§§1–65] was written on 20 loose sheets of lined foolscap, undated. These Wittgenstein left in his room in [Anscombe's] home in Oxford, where he lived...from April 1950 to February 1951. I (G.E.M.A.) am under the impression that he had written them in Vienna, where he stayed from the previous Christmas until March (Wittgenstein 1972: vi).

Parts II and III were written in Oxford, from April until September 1950. In February 1951, with his health rapidly deteriorating, Wittgenstein moved into the house of his doctor in Cambridge, Dr. Edward Bevan.[11] It was here that the final part, Part IV,

[9] At this point, Moore had fallen out of favor with the philosophical community. Ray Monk writes about this in *Prospect*, "He was the most revered philosopher of his era. So why did GE Moore disappear from history?" See https://www.prospectmagazine.co.uk/magazine/ge-moore-philosophy-books-analytic-ray-monk-biography

[10] This is evident to anyone who has read Moore's articles, especially "Proof of an External World," and then reads the opening passage of *On Certainty*.

[11] With only a couple months left to live, Wittgenstein told Dr. Bevan's wife, "I am going to work now as I have never worked before" (Malcolm 1962: 100). Indeed, during the last six weeks of his life, more than half of *On Certainty* (§§300–676) was written. These passages now makes up Part IV. His final entry, April 27, was just two days before he died.

was written—the first entry penned roughly a month and a half before his death.[12] Malcolm received a letter during this period in which Wittgenstein wrote:

> An extraordinary thing happened to me. About a month ago, I suddenly found myself in the right frame of mind for doing philosophy. I had been *absolutely* certain that I'd never again be able to do it. It's the first time after more than 2 years that the curtain in my brain has gone up.—Of course, so far I've only worked for about 5 weeks & it may be all over tomorrow; but it bucks me up a lot now…[A]part from a certain weakness which has constant ups & downs I'm feeling *very* well these days (Malcolm 1962: 99).

Less than two weeks later, Wittgenstein died of prostate cancer. He had turned 62 just three days prior.

## 1.4 Where *On Certainty* Fits in the Canon

Most scholars of Wittgenstein's philosophy divide his work into two distinct periods: the early and the later. In the early period, Wittgenstein was interested in depicting the world through language and logic, as demonstrated in his work, *Tractatus Logico-Philosophicus*. Perhaps the most important theory in this work is his *picture theory of meaning*. Language, he tells us, serves as "a *picture* of reality…a model of reality" (Wittgenstein 1967a: 4.01). The *picture theory of meaning* then, is a theory of linguistic representation, one that posits an isomorphic relationship, or one-to-one correspondence, between language and the world. If the picture is correctly depicted, we assign it a truth-value of *true*, if it is incorrect we assign it a truth-value of *false*.

In the later period, although his concentration remained on language, he was now interested in how we *use* it, as seen in his posthumously published work, *Philosophical Investigations*. His attention here was less focused on truth than on *meaning*. Early on in the work, Wittgenstein describes our acquisition of language like learning the rules of a game; in fact, he calls our use of language, "*language-games*." Although the term "game" can invite misunderstanding, implicitly bringing with it a lack of seriousness, even trivialization (e.g., "playing games," "gaming the system," "it's just a game," etc.), it might be more accurate to describe Wittgenstein's focus here on the *rules* within the language-game. "[T]he term 'language-game,'" he tells us, "is meant to bring into prominence the fact that... speaking [a] language is part of an activity" (Wittgenstein 1953: §23). This activity is measured *via* the rules existing in the game; every day, we engage in *numerous* language-games.[13] His emphasis here, in the *Investigations*, is on the wide

[12] In this final section, Wittgenstein recorded his dates of entry. The first one is dated March 10, 1951.

[13] As he says, "[r]eview the multiplicity of language-games in the following examples, and in others: Giving orders, and obeying them; Describing the appearance of an object, or giving its measurements; Constructing an object from a description (a drawing); Reporting an event; Speculating about an event; Forming and testing a hypothesis; Presenting the results of an experiment in tables and diagrams; Making up a story; and reading it; Play-acting; Singing catches; Guessing riddles;

variety of the uses of language, its elasticity, and its active role in a given community, or as he calls it, *form of life*.[14]

There are a growing number of people, however, Avrum Stroll, Daniele Moyal-Sharrock, Annalisa Coliva, myself, and others, who have made the case for a third, post-*Investigations*, period—a "Third Wittgenstein," as it were.[15] A third period of Wittgensteinian thought might elicit the obvious question: what sort of characteristics set *this* period apart from the other two? How is *it* different?

This third period of thought is different from the first two in at least three (related) ways. First, the notion of certainty, as Wittgenstein comes to understand it, is categorically distinct from knowledge; they differ in kind, not in degree. Second, this period exhibits a new kind of foundationalism, one where certainty "stands fast," outside and in support of, our many different language-games. Third, Common Sense truisms (what are sometimes called Wittgenstein's "hinge propositions") like "the earth exists" or "there are physical objects," are not empirical statements, they are not the sorts of things we learn about through experience, they are instead part of our "background," our "scaffolding," they help hold everything else in place.[16] These later transition from something "propositional" to something "actional." (In addition, in Chap. 6, I offer *another* difference, a further way in which certainty is acquired.)

While a third period of thought seems unmistakable, not everyone agrees that another period is necessary. William Child, in his book entitled *Wittgenstein*, says,

> similar themes appear in earlier work, including Wittgenstein 1937 *Notebooks* and in *Philosophical Investigations* sections 324–326 and 466–486. So *On Certainty* is not a completely new turn in Wittgenstein's philosophy, it develops thoughts that are already visible in his previous writings (Child 2011: 191).

---

Making a joke; and telling it; Solving a problem in practical arithmetic; Translating from one language into another; Asking, thanking, cursing, greeting, praying" (Wittgenstein 1953: §23).

[14] There is, of course, much more going on in both the *Tractatus* and the *Investigations*. There are a number of helpful guidebooks for both. For instance, for the *Tractatus*, there is G.E.M. Anscombe's, 1971 *An Introduction to Wittgenstein's Tractatus*. Philadelphia: University of Pennsylvania Press; Max Black's 1964 *A Companion to Wittgenstein's Tractatus*. Ithaca: Cornell University Press; Michael Morris' 2008 Routledge Philosophy Guide-Book to Wittgenstein and the Tractatus. New York: Routledge. For the Investigations, there is Garth Hallett's 1977 A Companion to Wittgenstein's Philosophical Investigations. Ithaca: Cornell University Press; Gordon Baker and P.M.S. Hacker's 2008 *Wittgenstein: Meaning and Mind (3 Volume Set of an Analytical Commentary on the Philosophical Investigations)*. Oxford: Wiley-Blackwell; David Pears' 1988 *The False Prison, Volume Two*, Oxford: Oxford University Press.

[15] This period includes several selections of notes that Wittgenstein did not intend for publication, such as, *Remarks on Color*, *Last Writings in the Philosophy of Psychology*, *Remarks on the Philosophy of Psychology*, Part II of *Philosophical Investigations*, and *On Certainty*.

[16] Moyal-Sharrock describes it this way: "The third Wittgenstein's conceptual rearrangements or re-categorizations (reshelvings) allow him to reshape the contours of epistemology and its contents: he points out our philosophical misuses of knowledge, highlights our confusion of knowledge with knowledge *claims*, and drives a categorical wedge between knowledge and foundational or primitive certainty which results in the latter's exclusion from epistemology and its redefinition as a *way of acting*" (Moyal-Sharrock 2004: 3).

And other scholars, like A.C. Grayling, believe that, while *On Certainty* is indeed different, it should be classified differently only because it exhibits a change in Wittgenstein's negative assessment towards philosophy:

> *On Certainty* appears to represent Wittgenstein's acceptance, at last, of philosophy's legitimacy as an enterprise…a refutation of scepticism and a characterization of knowledge and its justification (Grayling 2001: 305).

According to Grayling, then, because Wittgenstein finally focuses on traditional, philosophical problems, i.e., skepticism and knowledge, and tries to provide solutions to these problems, this marks Wittgenstein's departure from earlier work (early and later periods).

Grayling is right, but only to an extent. While Wittgenstein's work here is different, it is not due to his "acceptance…of philosophy's legitimacy." He does attempt a characterization of knowledge, but this, in turn, is what leads him to try to characterize *non-propositional certainty*. Characterizing non-propositional certainty seems to me to be quite *outside* the domain of epistemology; in fact, I'll be arguing that such a characterization is quite outside the philosophical mainstream, especially since Descartes.

While *On Certainty* contains new ideas regarding our notions of belief, knowledge, skepticism, and certainty, because the text consists largely of a series of working notes it is rough and sometimes difficult to decipher. As the editors tell us: this is "all first-draft material, which [Wittgenstein] did not live to excerpt and polish" (Wittgenstein 1972: vi). Avrum Stroll noted, "[m]any of the entries have the status of first thoughts, something to be put down on paper for further reflection or reconsideration" (Stroll 1994: 80). I think this is an important point to keep in mind while we read *On Certainty*.[17]

Elsewhere, I have described *On Certainty* as exploratory: an investigation of the philosophical landscape for purposes of discovery.[18] I still believe this is the case, however, it seems to me these notes also reveal an exploration of a different kind—an exploration into the practices and methodology of the discipline itself. Such an exploration can be uncomfortable in some respects, as it reveals flaws not only in philosophy, but also in the philosophers who willfully adhere to its traditional or conventional practices.

---

[17] I agree with Stroll who believes it is wrong to think of *On Certainty* as Wittgenstein's final say on these matters; however, there are others who dispute this idea, namely, Daniele Moyal-Sharrock. For more on this, see my 2014 book, *Exploring Certainty: Wittgenstein and Wide Fields of Thought*. Lanham, MD: Lexington Books, *xi*.

[18] Brice, Robert Greenleaf. 2014. *Exploring Certainty: Wittgenstein and Wide Fields of Thought*. Lanham: Lexington Books.

## 1.5 Resisting Our "Natural Biases"

Wittgenstein's conception of philosophy was narrowly circumscribed. For him, philosophy was limited to issues of epistemology, metaphysics, logic, and language. At one point in his book, *False Prison*, David Pears stops to ponder Wittgenstein's description of philosophy as "a battle against the bewitchment of our intelligence by means of language" (Wittgenstein 1953: §109). He concludes that Wittgenstein's "characterization of philosophy"

> is centered on our natural tendency to fall into certain kinds of error. Philosophy can only be understood as part of the pathology of the intellect. There is, of course, no suggestion that the error is willful: the point is that our minds have natural biases which philosophers must show us how to resist. But is there, perhaps, the further suggestion, that it is only these biases that create the need for philosophy, just as it is only mental illness that creates the need for psychotherapy? (Pears 1988: 216).

I agree with Pears' description of philosophy as "part of the pathology of the intellect." And while our "tendency to fall into certain kinds of error" may not be willful, I think it is fair to say that at least since Descartes, philosophy as a discipline, often now *intentionally*, *willfully*, pushes questions beyond the point of rationality. In this way, philosophy is complicit in creating biases that we must resist. This, I believe, is why Wittgenstein characterizes it as a "philosophical disease" (Wittgenstein 1953: §593), a "disease of thought" (Wittgenstein 1967a: §382) or sometimes a "disease of the understanding" (Wittgenstein 1978: §50). It is the willful spread of this disease that Wittgenstein says we now must confront.

Although it is philosophy that has permitted this disease to spread, paradoxically, it is only with the aid of a different kind of philosophy—Wittgenstein's therapeutic philosophy—that the hope for a cure might be delivered. In an odd way then, philosophy is both the illness *and* the cure. Consider what he says in different places throughout the corpus.

In *Philosophical Investigations*:

> §133: There is not *a* philosophical method, though there are indeed methods, like *different therapies*.
>
> §255: The philosopher's treatment of a question is like the treatment of an illness.

In *Zettel*:

> §382: In philosophizing we may not *terminate* a disease of thought. It must run its natural course, and *slow* cure is all important (Wittgenstein 1967b: §382).

And in his *Big Typescript*:

> §407: [W]ork on philosophy is actually closer to working on oneself. On one's own understanding. On the way one sees things. (And on what one demands of them.) (Wittgenstein 2012: §407).

As such, it is important to distinguish, early on in our analysis, the two kinds of philosophy he is discussing here. There is, on the one hand, the sickness—toxic Cartesian skepticism, while on the other, there is the cure—Wittgenstein's therapeutic Common Sense Approach. Throughout the text, we will be looking at where and how philosophy shifts from sickness to cure.

## 1.6 Wittgenstein's Therapeutic Approach & Scholarship

For years, notable scholars, like P.M.S. Hacker, have argued that Wittgenstein's "conception of philosophy is predominately therapeutic" (Hacker 1972: 116).[19] Few scholars dispute this. When analyzed through a therapeutic lens, however, this scholarship has focused almost exclusively on *Freudian* psychoanalysis. This is not unreasonable given Wittgenstein's admiration for Freud.[20] Yet, in light of the success and effectiveness of other psychotherapies, I believe such exclusive focus is shortsighted.[21] Furthermore, Wittgenstein encouraged us to seek a broader perspective; there are, as he says, *different* philosophical methods "*like different therapies*" (Wittgenstein 1953: §133).

A reasonable place to begin an investigation into scholarship of Wittgenstein's therapeutic approach is with *The New Wittgenstein*. In this collection, Cora Diamond, James Conant and others argue for the continuity of a therapeutic conception of philosophy in Wittgenstein's thought, tracing it from his early to his later works. The problem, however, is that in surprisingly few instances do we find anyone extending his therapeutic conception beyond work in the *Investigations* to *On Certainty*.

Alternatively, in his book, *Wittgenstein and On Certainty*, Andy Hamilton says that while "[i]t is true that *On Certainty* does not craft a therapeutic approach in the manner of the *Investigations*" (Hamilton 2014: 269), nonetheless, "the therapeutic approach is fully present in *On Certainty*" (Hamilton 2014: 11).

> The open quality of his less crafted writings such as *On Certainty* reflects Wittgenstein's essentially therapeutic method...Many of his claims have an air of the provisional or tentative—a sense that both author and reader must continue to wrestle with the issue (Hamilton 2014: 55).

"Wrestling with issues" nicely illustrates the internal struggle that occurs with this kind of therapy. Hamilton describes this approach to philosophical ideas as "combative," an approach that encourages the patient to occasionally grapple with contrary positions—*simultaneously*—in the mind.[22] This inner struggle "can be therapeutic," says Hamilton, but, more importantly, "wrestling with issues" illustrates that "combative therapy…is…*cognitive*" (Hamilton 2014: 266); it forces the patient to *try* to make sense of two equally reasonable, but contrary, positions.

"Combative therapy" is set in contradistinction to another way scholars have interpreted Wittgenstein's therapeutic approach, a method known as *quietism*.

[19] Hacker disagrees with Diamond, Conant, and others who, in the *New Wittgenstein*, argue for the continuity of a therapeutic conception of philosophy in Wittgenstein's thought from the early to the later works. His "dissenting voice" serves as the final chapter in the work, entitled, "Was he trying to whistle it?" (Hacker 2000: 353–388).

[20] At one point, Wittgenstein even described himself as "a disciple of Freud" (Wittgenstein 1967c: 41).

[21] I discuss this in subsequent chapters in more detail.

[22] Note, this is something traditional philosophy would dismiss immediately.

*Quietism* has its origins in an ancient form of skepticism known as *Pyrrhonism*. This form of skepticism takes a rather cynical approach toward philosophy as a discipline, insisting that, because we cannot ascertain how things truly are in the world, we must recognize our lack of comprehension (ακαταλεπσια) and "suspend judgment." Suspending judgment, it is argued, will return us to a quieter point, before we were "disturbed by philosophical questions" (Gutshmidt 2020: 105).

Associating Wittgenstein's therapeutic approach with *this* form of skepticism, however, is, I believe, to *misread* Wittgenstein. As Hamilton correctly points out: "Wittgenstein's philosophical position…aim[s] to refute traditional [i.e., *Cartesian*] skepticism," *not* refute philosophy (Hamilton 2014: 267). "Finding peace," Wittgenstein tells us, is the goal (Wittgenstein 1984: 43e), but nowhere does he suggest this peace ought to come at the expense of refuting the entire discipline of philosophy. Rather, for Wittgenstein, "finding peace" will be accomplished by *reshaping* contrary points of view; "finding peace" will be accomplished by *recognizing* and respecting a Common Sense scope and limit to skeptical queries. Such actions call for careful thought; hence, such actions are *cognitive*. But *quietism*, as Hamilton says, is "non-cognitive." Overcoming intellectual philosophical problems means one must struggle, wrestle, grapple with them, which makes *quietism* an "unacceptably passive notion of therapy" (Hamilton 2014: 266).

I agree with much of what Hamilton has to say about Wittgenstein's therapeutic approach to *On Certainty*. I think his recognition that there is a therapeutic method present in the work, that there is a cognitive aspect to this therapy, and that this cognitive aspect requires actively (not passively) addressing the problem, is correct. Yet, Hamilton's greatest contribution to a fuller understanding of a therapeutic method in *On Certainty* is one he never explicitly makes. Hamilton does not take the next step of *connecting* Wittgenstein's therapeutic method to an *actual* therapy. I believe an investigation into such a connection could prove to be enlightening. It therefore seems reasonable to consider such a connection, which is what I have set out to do in this book.

In the following chapter (Chap. 2), I will offer a broader explication of philosophical therapy before aligning it with a modern psychotherapy known as *Cognitive Behavioral Therapy* (*CBT*).

## References


Anscombe, G.E.M. 1971. *An Introduction to Wittgenstein's Tractatus*. Philadelphia: University of Pennsylvania Press.

Aquinas, St. Thomas. 1981. (Originally published 1485.) *Summa Theologiae of St. Thomas Aquinas* Vol. 1, Pt. I, Quest. 1, Article 4. Notre Dame: Christian Classics; English Dominican Province Translation edition.

Baker, Gordon, and P.M.S. Hacker. 2008. *Wittgenstein: Meaning and Mind (3 Volume Set of an Analytical Commentary on the Philosophical Investigations)*. Oxford: Wiley-Blackwell.

Black, Max. 1964. *A Companion to Wittgenstein's Tractatus*. Ithaca: Cornell University Press.

Brice, Robert Greenleaf. 2014. *Exploring Certainty: Wittgenstein and Wide Fields of Thought*. Lanham: Lexington Books.

Child, William. 2011. *Wittgenstein*. New York: Routledge Press.

Conway, Gertrude. 1989. *Wittgenstein On Foundations*. Atlantic Highlands: Humanities Press International.

Descartes, Rene. 1993. (Originally published 1641.) *Meditations on First Philosophy*, 3rd ed, trans. Donald Cress. Indianapolis: Hackett Publishing.

Frongia, G., and B. McGuiness. 1990. *Wittgenstein: A Bibliographical Guide*. Oxford: Basil Blackwell Press.

Grayling, A.C. 2001. Wittgenstein on Scepticism and Certainty. In *Wittgenstein: A Critical Reader*, ed. Hans Johann Glock. Malden: Blackwell.

Gutshmidt, Rico. 2020. Beyond Quietism: Transformative Experience in Pyrrhonism and Wittgenstein. *The International Journal for the Study of Skepticism* 10 (2): 105–128.

Hacker, P.M.S. 1972. *Insight and Illusion: Wittgenstein on Philosophy and the Metaphysics of Experience*. Oxford: Oxford University Press.

———. 2000. Was he trying to whistle it? In *The New Wittgenstein*, ed. Crary and Read. New York: Routledge Press.

Hallett, Garth. 1977. *A Companion to Wittgenstein's Philosophical Investigations*. Ithaca: Cornell University Press.

Hamilton, Andy. 2014. *Wittgenstein and On Certainty*. New York: Routledge Press.

Hume, David. 2007. (Originally published 1748.) *Enquiry Concerning Human Understanding* Section VIII, "Of Liberty and Necessity," Oxford: Oxford University Press.

Leibniz, Gottfried. 1982. (Originally published 1686.) *Generales Inquisitiones De Analysi Notionum et Veritatum*, vol. 388, ed. Franz Schupp. Leipzig: Meiner Publishing.

Locke, John. 1975. (Originally published 1689.) *An Essay Concerning Human Understanding*. Oxford: Clarendon Press.

Malcolm, Norman. 1962. *Ludwig Wittgenstein: A Memoir*. London: Oxford University Press.

Malgren, Helge. 1971. *Intentionality and Knowledge: Studies in the Philosophy of G.E. Moore and Ludwig Wittgenstein*. Goteborg: Goteborgs Universitet.

McGinn, Marie. 1989. *Sense and Certainty*. Oxford: Oxford University Press.

Moore, G.E. 1925. Defense of Common Sense. Reprinted in *Philosophical Papers*. New York: Collier Books, 1962.

Morris, Michael. 2008. *Routledge Philosophy Guide-Book to Wittgenstein and the Tractatus*. New York: Routledge.

Moyal-Sharrock, Danielle. 2004. *The Third Wittgenstein: Post-Investigations Works*. Burlington: Ashgate Publishing.

Pears, David. 1988. *The False Prison, Volume II*. Oxford: Oxford University Press.

Russell, Bertrand. 1912. *Problems of Philosophy*. Oxford: Oxford University Press.

Stroll, Avrum. 1994. *Moore and Wittgenstein on Certainty*. Oxford: Oxford University Press.

Svensson, Gunnar. 1981. *On Doubting the Reality of Reality: Moore and Wittgenstein on Sceptical Doubts*. Stockholm: Almqvist & Wiksell International.

Wilde, Carolyn. 1976. *Certainty: A Discussion of Wittgenstein's Notes in On Certainty*. London: Open University Press.

Wittgenstein, Ludwig. 1953. *Philosophical Investigations*, trans. G.E.M. Anscombe. Oxford: Blackwell Publishers.

———. 1967a. (Originally published 1922.) *Tractatus Logico-Philosophicus*, trans. David Pears and Brian McGuinness. New York: Routledge.

———. 1967b. *Zettel*, 2nd ed, eds. G.E.M. Anscombe and G.H. von Wright, trans. G.E.M. Anscombe. Oxford: Blackwell Publishers Ltd.

———. 1967c. *Lectures and Conversations on Aesthetics, Psychology, and Religious Belief*, ed. C. Barrett. Berkeley: University of California Press.

———. 1972. *On Certainty*, eds. G.E.M. Anscombe and G.H. Von Wright, trans. D. Paul and G.E.M. Anscombe. New York: Harper & Row.

———. 1984. *Culture and Value*, eds. G.E.M. Anscombe and G.H. von Wright, trans. Peter Winch. Chicago: University of Chicago Press.
———. 2012. *Big Typescript*, eds. and trans. C. Grant Luckhardt and Maximilian A.E. Aue. Oxford: Wiley-Blackwell Publishers.
Wolgast, Elizabeth. 1977. *Paradoxes of Knowledge*. Ithaca: Cornell University Press.

# Chapter 2
# Introduction: Philosophical Therapy

**Abstract** This chapter provides a brief introduction to the notion of philosophical therapy, its connection to the eighteenth century *Scottish School of Common Sense*, and further back still to its connection with ancient Greek philosophy.

**Keywords** Philosophical therapy · Cognitive Behavioral Therapy (CBT) · Common sense · Psychotherapy · James Braid · Therapeutic · Rene Descartes · Ludwig Wittgenstein · Aaron Beck · Albert Ellis

## 2.1 What Is Philosophical Therapy?

Broadly speaking, *philosophical therapy* attempts to treat philosophical confusions with psychotherapeutic tools. It aims at fostering a new way of thinking about what appear to be intractable philosophical problems. Despite how unusual it may sound, the combination of these two disciplines is not new.

In Pierre Hadot's important study of ancient Greek philosophy, he argues that, for the ancient Greeks, philosophy was not strictly an *intellectual* pursuit but a means for achieving happiness by transforming one's perception of, and being in, the world. This "complex interrelation" between critical discourse and a "way of living and of seeing the world," says Hadot,

> corresponds, above all, to the choice of a certain way of life and existential option which demands from the individual a total change of lifestyle, a conversion of one's entire being, and ultimately a certain desire to be and to live in a certain way (Hadot 2004: 3).

Similarly, as Donald Robertson recently pointed out,

> Philosophy, to a large extent, has always been about transforming the life of the philosopher, in a manner broadly resembling modern psychotherapy…As far back as the Socrates of Plato's *Gorgias*, philosophy has been compared to the art of medicine for *psyche*, the mind or soul. In other words, what we now call "psychotherapy" was explicitly recognized as an aspect of philosophy in ancient Greece and Rome (Robertson 2019: 1).



R. G. Brice, *Wittgenstein's On Certainty: Insight and Method*, SpringerBriefs in Philosophy, https://doi.org/10.1007/978-3-030-90781-5_2

Focusing mostly on the ancient philosophical traditions of the Stoics, Robertson considers their relationship to a modern psychotherapy known as *Cognitive Behavioral Therapy (CBT)*.

According to one of its pioneers, Aaron Beck,[1] CBT is "an approach to treatment designed to identify, evaluate, and change…maladaptive belief systems and dysfunctional styles of information processing" (Beck et al. 1979: 4). CBT is a method that seeks to help the patient "deal with irrational or disturbing emotions" in an effort to, as Robertson puts it, "cultivate rational, healthy, and proportionate [emotions] in their stead" (Robertson 2019: 15). Cultivating these emotions is most successful when the patient, with the assistance of the cognitive therapist, commits to the following strategies:

1. monitoring negative automatic thoughts, or cognitions.
2. evaluating the relationship between thoughts, feelings, and actions.
3. carefully evaluating the evidence for and against distorted or maladaptive cognitions.
4. generating alternative cognitions and to substitute them for the negative ones.
5. identifying and modifying underlying dysfunctional assumptions and beliefs which predispose one to negative automatic thoughts (Beck et al. 1979: 4).

These strategies, says Tim O'Keefe, have made CBT "more effective than alternative treatments" (O'Keefe 2020: 429). In particular, CBT has become one of the most relied upon therapeutic approaches for anxiety and depression. A meta-analytic review found that

> [t]he superiority of CBT over alternative therapies was evident…among patients with anxiety or depressive disorders. These results…suggest that CBT should be considered a first-line psychosocial treatment of choice…for patients with anxiety and depressive disorders (Tolin 2010: 710).

CBT is, as Robertson correctly points out, "the predominant school of modern evidence-based psychological therapy" (Robertson 2019: 15).

## 2.2 Tracing the Roots of Cognitive Behavioral Therapy (CBT)

CBT has its roots in "various schools of classical philosophy." As Albert Ellis, another founder of the theory explains, "[m]any of the principles incorporated in the theory…are not new; some of them, in fact, were originally stated several thousand years ago, especially by the Greek and Roman Stoic philosophers" (Ellis 1962: 35). Beck too cites the Greeks, insisting that his "cognitive therapy uses primarily the

[1] In the mid-1960s, as a response to what they saw as the ineffectiveness of Freudian psychoanalysis, Aaron Beck and Albert Ellis began, independently, developing empirically driven therapies. Beck researched what he called *Cognitive Therapy* (*CT*) and *Cognitive-Behavioral Therapy* (*CBT*), while Ellis developed what came to be known as *Rational Emotive Behavior Therapy* (*REBT*). They later worked together.

Socratic method" (Beck et al. 2005: 167).[2] Similarly, O'Keefe draws connections between CBT and another Greek philosopher, Epicurus.[3] Epicurus recommends we "acquire self-knowledge," in order to, as O'Keefe says, "examine our beliefs, desires, and ways of life" (O'Keefe 2020: 429). This kind of self-reflective examination

> is similar to cognitive-behavioral therapy, whose 'defining feature…is the proposition that symptoms and dysfunctional behaviors are often cognitively mediated and, hence, improvement can be produced by modifying dysfunctional thinking and beliefs'" (O'Keefe 2020: 429).

By blending ancient Greek philosophical methods with contemporary psychotherapeutic strategies, CBT can develop and reinforce rational and healthy emotions in the patient by modifying and reshaping irrational and dysfunctional feelings. This mix between philosophical methods and therapeutic strategies can be quite successful when it remains balanced. The problem, however, is that it has not always remained that way.

In the intervening years since Socrates, Epicurus, and the Stoics, philosophical themes and ideas have vacillated. Sometimes they tack toward the ancient Greek goal of "transforming the life of the philosopher,"[4] other times, however, they do not.[5] In the seventeenth century, philosophical inquiry had moved so far away from the ancient Greek goal that thinkers like Rene Descartes sought to intellectually question[6] *everything* in an effort to arrive at something that *could not be doubted*. As long as it is *logically* possible for a proposition to be true (that is, even though wildly improbable, not self-contradictory) it was permitted to stand, no matter how impractical. Philosophical methods had become detached from human reality. Rather than transforming lives, philosophy *itself* had transformed, becoming some-

---

[2] However, in an earlier work Beck, like Ellis, says the "philosophical origins of cognitive therapy can be traced back to the Stoic philosophers" (Beck, Rush, Shaw, Emery.1979: 8).

[3] Robertson makes this connection too.

[4] In Sharon Lebell's interpretive translation of the Stoic philosopher, Epictetus, entitled, *The Art of Living: the Classical Manual on Virtue, Happiness, and Effectiveness*, Lebell writes, "Epictetus would have had little patience for the aggressive position-taking and -defending and verbal pirouettes that unfortunately sometimes pass for 'doing' philosophy in today's universities…Inasmuch as he passionately denounced displays of cleverness for its own sake, he was committed to non-patronizing explanations of helpful ideas for living well. He considered himself successful when his ideas were easily grasped and put to use in someone's real life, where they could actually do some good elevating that person's character" (Lebell 2007: vvii).

[5] For instance, the Cynics and the Skeptics, the latter of which was still active in the third century CE, especially in Plato's Academy, both were tied to the goal of a way of life free from anxieties as well as the goal of intellectual "peace of soul." My thanks to Paul Streveler for pointing this out to me.

[6] I say "intellectually question" in order to emphasize how Cartesian or intellectualist philosophy does *not* take seriously the second of the 5 CBT strategies listed above; particularly that our thinking and emotions have to square with our actions. Radical skepticism is so inconsistent with our actions that no one can *live a life* as a radical skeptic. It's a purely notional or parlor game conception.

thing of an intellectualist[7] game of focusing on logical *possibilities* rather than on empirical and practical *probabilities*. These methods and strategies did succeed in reuniting again in the late-eighteenth and early-nineteenth centuries.

Standing in opposition to radical skepticism, Thomas Reid founded his Scottish School of Common Sense Realism.[8] Appealing to the sorts of basic principles that no reasonable person could possibly refute, Reid insisted that,

> [t]he evidence of sense, the evidence of memory, and the evidence of the necessary relations of things, are all distinct and original kinds of evidence, equally grounded on our constitution…To reason against any of these kinds of evidence is absurd…They are first principles; and such fall not within the province of reason, but of common sense (Reid 1915: 44–45).

These "first principles" were an appealing alternative to Cartesian skepticism, and Reid's School succeeded in attracting many distinguished scientists and academics. These students of Common Sense sought to establish observation and experimentation—the main pillars of empirical science—over the exhaustive, and often fruitless, intellectualist inquiry that Descartes' methodological skepticism favored. Some of the better-known followers of the School included, Dugald Stewart, Thomas Brown, and the surgeon, James Braid.[9] It was Braid, in particular, who,

[7] A distinction should be made here between "intellectual" and "intellectualist." An intellectual is someone who, after receiving a thorough education, possesses a highly developed intellect; this is a good and highly sought after goal. But being *intellectualist*, on the other hand, is an excessive devotion to the intellect. By way of comparison, think of the difference between scientific and scientistic, moral and moralistic.

[8] It should be noted that Reid's appeal to Common Sense developed mostly out of his frustration with *Humean* skepticism. Broadly speaking, the seventeenth and eighteenth centuries produced two veins of skepticism: Cartesian skepticism that which emerged as a result of David Hume's *Problem of Induction*. Descartes questioned whether one could trust one's senses, whether one was dreaming, and finally, whether one lived in a world where an evil demon was trying to deceive people on things they found most manifest. The sort of skepticism David Hume introduced—the *Problem of Induction*—stemmed from his analysis of inductive reasoning. To reason inductively means one begins with a limited sampling of something before arriving at a generalization based on that limited collection. For example, the sun came up this morning. The sun came up the morning prior, too. In fact, as far back as I can remember, the sun has risen. Hume says that "[f]rom causes which appear *similar* we expect similar effects." Therefore, I should expect a similar effect with the sun tomorrow. Here, I am making an inference that the sun will come up tomorrow just as it has in the past. In fact, *all inference* from past and present experience rests on the very *assumption* that the future will resemble the past. But Hume noticed that this leads to the extremely skeptical conclusion that much of what we *think* we know we really *don't* know. For our assumption that the future will resemble the past is itself an unprovable assumption. How do we really *know* that the future will resemble the past? We can't say that it's because experience shows us that the future has always resembled the past. For a moment's reflection illustrates that our experience *comes from past events*. That is, one cannot say that the future will resemble the past because *the future has always resembled the past*, for now one is appealing to the very principle that we wish to establish, and note, one is appealing to it *through induction*.

[9] Braid was not an academic philosopher but he *was* influenced by the philosophers of his day. He cites and references the "Scottish realist philosophers" throughout his writing.

after attending a demonstration on hypnotism,[10] helped redirect philosophy back to the ancient Greek goal to transform the life, practical as well as intellectual, of the philosopher.

Braid, recognizing the therapeutic possibilities of hypnotism, ultimately succeeded in recreating its "trance-like" state by relying on the scientific tools and techniques promoted by Common Sense Realists.[11] Robertson describes Braid's contribution this way:

> Braid…attempted to take the therapeutic practices of mesmerism [or hypnosis] and reinterpret them in light of Scottish realist ("common sense") philosophy of mind, substituting the psychological laws of association, habit, sympathy, and suggestion, etc, for the supernatural theory of "animal magnetism" (Robertson 2019: 3).

While the hypnotism connection is somewhat unexpected, the important connection here is Braid's help in re-establishing philosophical methods and therapeutic strategies.[12] As Braid himself put it:

---

[10] Hypnotism has had a few different names. Initially it was called "animal magnetism," a theory developed by Franz Mesmer (1734–1815). Mesmer believed a magnetic force existed throughout the universe. This force affected the health of the human body. He experimented with magnets in an effort to heal sick bodies. While grasping magnets, Mesmer slowly waved his hands before people. "[A]round 1774 he…[began] referring to this action as making 'Mesmeric passes.' He used the word 'mesmerize,' formed from his last name" https://www.bookofdaystales.com/hypnosis/ After James Braid began applying scientific tools to it, he was able to reproduce the state without any of the "occult" connections that Mesmer had associated with it. Braid renamed this state, "hypnosis" (from the Greek *hypno,* meaning "sleep"). He described hypnosis as a "shift of the nervous system into a new condition" (Tinterow 1970: 271).

[11] In his 1843 book, *Neurypnology*, Braid described it this way: "I now proceed to detail the mode which I practise for inducing the phenomena. Take any bright object (I generally use my lancet case) between the thumb and fore and middle fingers of the left hand; hold it from about eight to fifteen inches from the eyes, at such position above the forehead is may be necessary to produce the greatest possible strain upon the eyes and eyelids, and enable the patient to maintain a steady fixed stare at the object. The patient must be made to understand that he is to keep the eyes steadily fixed on the object, and the mind riveted on the idea of that one object. It will be observed, that owing to the consensual adjustment of the eyes, the pupils will be at first contracted: they will shortly begin to dilate, and after they have done so to a considerable extent, and have assumed a wavy motion, if the fore and middle fingers of the right hand, extended and a little separated, are carried from the object towards the eyes, most probably the eyelids will close involuntarily, with a vibratory motion. If this is not the case, or the patient allows the *eyeballs to move*, desire him to begin anew, giving him to understand that he is to allow the eyelids to close when the fingers are again carried towards the eyes, but that the eyeballs *must be kept fixed , in the same position, and the mind riveted to the one idea of the object held above the eyes*" (Braid 2010: 27–28).

[12] And although these philosophical methods can be traced back to the ancient Greeks, Braid grounds them in Common Sense. Some scholars argue a direct line can be drawn back to Aristotle from the Common Sense Realists. For instance, in Bernard Mahoney's *The Empirical Tradition and Newman's Concept of the Conscience* (his Ph.D. dissertation), Mahoney contends that the line of British Empiricists—Locke, Hume, and to a certain degree Berkeley—begins with the study of Aristotle at Oxford. When Aristotle was banned at the University of Paris in the thirteenth century, a group of scholars moved north to Oxford to continue their research. These scholars, inspired by the empiricism of Aristotle, influenced not only the aforementioned philosophers but also (much later) exercised an amount of influence on Thomas Reid and his Scottish School of Common

> [i]t was not…the induction of the *nervous sleep* [hypnotism] alone which effected the cure, but my knowledge of how to *direct the influence* DURING the sleep, so as to break down the pre-existing, involuntary fixed, dominant idea in the patient's mind, and its consequences (Robertson 2013: 97).

In other words, while hypnotism was the tool Braid used to arrive at the patient's underlying problem/s, it was his application of *philosophical methods*—those "first principles" the Scottish School of Common Sense was founded upon—that helped direct his knowledge. For this reason, Braid deserves recognition for reuniting philosophical methods together with therapeutic strategies. Indeed, Robertson, in his *Discovery of Hypnosis*, believes

> Braid has a right to be seen, at least in some respects, as a distant forerunner of modern cognitive behavioral therapists. In addition to defining hypnotic therapy in terms of specific cognitive behavioral interventions, he clearly assumed a cognitive theory of psychological disorders and based his whole methodology upon a rational, empirical, scientific and Common Sense orientation, of the kind espoused in modern CBT (Robertson 2013: 17).

Though less an emphasis on the *School* of Common Sense than our common sensibilities, the connection between common sense and therapy is also echoed in Beck's writings. In *Cognitive Therapy and the Emotional Disorders*, Beck devotes an entire chapter to the important role common sense plays in developing modern cognitive therapy. At one point, lamenting the fact that some of his colleagues have a "tendency to ignore the importance of common sense psychology," Beck, quoting G. Allport, poses the following question:

> How in the helping professions—and here I include psychiatry, the ministry, social work, applied psychology, and education—can we recover some of the common sense that we seem to have lost along the way? (Beck 1979: 9–10).

It seems the Common Sense recovery efforts of the 1960s therapeutic community proved remarkably successful. By reuniting philosophical methods with therapeutic strategies, the important work of Thomas Reid and others in the School of Common Sense who influenced those who came after them, can be seen more clearly. The concern confronting us now, as we move forward, is how to go about *reinforcing* the connection forged between our philosophical methods and these therapeutic strategies.

---

Sense. Reid's School is an important link to the empirical tradition, for it can be seen in the work of Henry Sidgwick, G.E. Moore's instructor at Cambridge. Reid's work can also be seen in a contemporary of Moore's, H. H. Price, who, in his book *Perception*, said, "[t]he position maintained in this chapter with regard to the nature and validity of perceptual consciousness is in essence identical with that maintained by Reid against Hume" (Price 1984 203.) Mahoney's study shows that the British empirical tradition began at Oxford well before Locke, Berkeley, and Hume, and, that these Aristotelian scholars have influenced every empirical movement well into the twentieth century including the School of Ordinary Language, i.e., Wittgenstein, J. L. Austin.

## 2.3 Possible Criticisms & Replies

Of course, it could be argued that what philosophy has become is a natural outgrowth of a *kind* of intellectualism, one directly linked to the Socratic method. Socrates, after all, engaged in a discursive back and forth, often using skeptical queries to trap unsuspecting Athenians into logical contradictions. Descartes merely carries this tradition to its logical conclusion. How then is Descartes different from Socrates?

The key difference between Descartes and Socrates is that Socrates believed in his method to such a degree that he was willing to die for it—a sacrifice that would imply he wished to *transform lives*. "Life without examination," Socrates famously declares, "is not worth living" (Plato 1961: St. 38a). For Socrates, then, philosophy must make a difference to the way we actually lead our lives. We do not see this in the methodological skepticism championed by Descartes. What we *do* see, however, is movement *away* from *practical* philosophical growth, and movement *toward* impractical, intellectualist games. For this and other reasons, I think it is fair to say that Descartes is less revered now than in the past.

Despite his diminished influence, the dysfunction caused by Descartes' framework and methodology can still be felt in contemporary philosophy, in particular, those philosophers committed to certain kinds of realism. Consider, for instance, Stuart Hampshire's belief that we will, at some point, arrive at eternal ethical truths. In his book, *Morality and Conflict*, Hampshire presents the "doctrine of moral harmony," a single harmonious scheme of morality (Hampshire 1983: 144). The idea is that as we acquire more knowledge, as we improve our reasoning, we will eventually be able to agree on *one set of rules and principles*. This doctrine of moral harmony, says Hampshire, will thus provide us with a single right answer for our moral questions. The problem, however, is that freedom and openness of thought will likely foster *new* possibilities—possibilities we cannot currently imagine. Because of this, it seems reasonable to assume that our moral rules will not *converge* but *diverge*.

Another criticism here might be that by using some of the strategies developed *in* the discipline to analyze the discipline *itself*, this approach proves circular. However, this seems problematic for two reasons. First, it trivializes philosophy, once again reducing it to an exploration of mere logical possibilities instead of promoting it as a discipline that has the potential to "transform lives." Second, although the strategies found in CBT may have their *origins* in ancient philosophy, this does not mean they cannot be used *on* philosophy itself. The fact that certain tools are forged from a particular kind of metal doesn't mean the tools cannot be used on the metal itself. Indeed, the strategies found in CBT will serve as vital tools in our philosophical toolkit—we will use these instruments in an effort to adjust and reshape irrational thoughts and beliefs that emerge (and re-emerge) in contemporary philosophy.

A third criticism, somewhat related to the one above, is that philosophy is both the disease *and* the cure. However, the combination of logical possibility and radical skepticism—characteristic of Cartesianism—*willfully* brought this on. It is now up to Wittgenstein's *therapeutic philosophy* to deliver the cure.

## 2.4 Challenges of the Method

While the analogies between CBT and Cartesian skeptics are numerous, it might be worthwhile to attend to a *dis*analogy—especially the one between the non-functioning patient of CBT who realizes she's suffering from, say, disabling anxiety, and the professional philosopher who resolutely dismisses the suggestion that his philosophical outlook and methodology is cognitively dysfunctional.

Although the person suffering from severe anxiety needs the help of (and may, of her own volition, actively seek out) the CBT professional, the professional philosopher, on the other hand, when confronted with errors, may flat out deny that any procedural errors exist, and, in turn, reject any help. From a psychological perspective, it might be easy to write off the philosopher as being in a state of denial, but this doesn't help him confront the issue. Nor does it treat the problem. Reaching this person can be difficult, sometimes impossible. However, by appealing to common sense, by petitioning his practical sensibilities, the hope is that he will at least consider an alternative to the Cartesian framework. Indeed, it is only with such philosophers that we can hope to make any reasonable progress.

Therapeutic philosophy is a challenging method.[13] It involves struggle, occasional confusion, false starts, and a willingness to stop and begin all over again. It requires resilience in the face of intellectually uncomfortable truths. Recognizing a personal belief is mistaken can be difficult to admit, especially to one's self. As Wittgenstein says in *Culture and Value*, "[n]othing is so difficult as *not* deceiving oneself" (Wittgenstein 1984: 34e, my emphasis). It takes strength and courage to admit that some (or even most) of your philosophical beliefs are flawed or wrong. It also requires determination. Once committed to this method, it is reasonable to assume the process will take some time before one emerges a less troubled, less obsessed philosopher. As Wittgenstein puts it: "[i]n philosophizing we may not *terminate* a disease of thought. It must run its natural course and *slow* cure is all important" (Wittgenstein 1967: §382). This guideline, however, elicits an important question, namely, is this philosophical disease—radical skepticism—*curable*? Can Wittgenstein's therapeutic philosophy help prevent this "disease of thought"?

Throughout the course of this book, I intend to show how therapeutic philosophy not only *can* help us, it *must*. Radical skepticism *is* curable, but like any widespread disease, before we can cure it, we must *contain* it. And like any good therapy, there is not an endpoint; it is ongoing. *Philosophical therapy* requires continuous work.

[13] Engaging in philosophy was often deeply personal for Wittgenstein. Indeed, it seems Wittgenstein was sometimes unable to separate the two, as Bertrand Russell recounts: "He used to come to see me every evening at midnight, and pace up and down the room like a wild beast for three hours in agitated silence. Once I said to him: 'Are you thinking about logic, or about your sins?' 'Both,' he replied, and continued his pacing. I did not like to suggest it was time for bed, for it seemed probable both to him and to me that on leaving me he would commit suicide" (Russell 1968: 137).

In the following chapter (Chap. 3), I devote some time to discussing those whose writings played a role in shaping Wittgenstein's thoughts for *On Certainty*. These primarily include, Rene Descartes (although never mentioned by name) and G.E. Moore.

## References


Beck, Aaron. 1979. *Cognitive Therapy and the Emotional Disorders*. New York: Penguin Books.

Beck, Aaron, John Rush, Brian Shaw, and Gary Emery. 1979. *Cognitive Therapy of Depression*. New York: Guilford.

Beck, Aaron, Gary Emery, and R. Greenberg. 2005. *Anxiety Disorders and Phobias: A Cognitive Perspective*. Cambridge: Basic Books.

Braid, James. 1845. On Hypnotism. *The Lancet* 1: 627–628.

———. *Neurypnology*. 1843. Republished in 2010. London: Kessinger Publishing Company.

Ellis, Albert. 1962. *Reason and Emotion in Psychotherapy: A Comprehensive Method of Treating Human Disturbance*. Secaucus: Citadel.

Hadot, Pierre. 2004. *What Is Ancient Philosophy?* Cambridge: Belknap Press.

Hampshire, Stuart. 1983. *Morality and Conflict*. Cambridge: Harvard University Press.

Lebell, Sharon. 2007. *The Art of Living: the Classical Manual on Virtue, Happiness, and Effectiveness*. New York: Harper One.

O'Keefe, Tim. 2020. Epicurean Advice for the Modern Consumer. In *The Routledge Handbook of Hellenistic Philosophy*, ed. Kelly Arenson. Abingdon: Routledge.

Plato. 1961. *The Collected Dialogues of Plato*. Princeton: Princeton University Press.

Price, H.H. 1984. *Perception*. Westport: Greenwood Publishing Group.

Reid, Thomas. 2012. *Selections from the Scottish Philosophy of Common Sense*. Veritatis Splendor Publications. Originally published, Chicago: Open Court, 1915.

Robertson, Donald. 2013. *The Discovery of Hypnosis: The Complete Writings of James Braid, the Father of Hypnotherapy*. Lulu.com.

———. 2019. *The Philosophy of Cognitive-Behavioural Therapy (CBT): Stoic Philosophy as Rational and Cognitive Psychotherapy*. New York: Routledge.

Russell, Bertrand. 1968. *The Autobiography of Bertrand Russell*. Vol. II. New York: Allen & Unwin Publishing.

Tinterow, Maurice. 1970. *Foundations of Hypnosis: From Mesmer to Freud*. Springfield: Charles C. Thomas.

Tolin, David. 2010. Is Cognitive–Behavioral Therapy More Effective Than Other Therapies: A Meta-Analytic Review. *Clinical Psychology Review* 30 (6): 710–720.

Wittgenstein, Ludwig. 1967. *Zettel*, 2nd ed, eds. G.E.M. Anscombe and G.H. von Wright, trans. G.E.M. Anscombe. Oxford: Blackwell Publishers Ltd.

———. 1984. *Culture and Value*, eds. G.E.M. Anscombe and G.H. von Wright, trans. Peter Winch. Chicago: University of Chicago Press.

# Chapter 3
# Background to *On Certainty*

**Abstract** This chapter focuses on three related issues: (a) the extreme form of skepticism developed by Rene Descartes, (b) G.E. Moore's Common Sense response to it, and (c) Wittgenstein's reply to both.

**Keywords** Common Sense · Proof · Rene Descartes · Knowledge · G.E. Moore · Psychotherapy · Reality · Skepticism · Therapeutic · Ludwig Wittgenstein

## 3.1 Descartes, Moore, & Wittgenstein[1]

In an effort to "find just one thing, however slight, that is certain and unshaken," seventeenth century philosopher, Rene Descartes, questioned anything that could possibly be doubted:

> I suppose that everything I see is false. I believe that none of what my deceitful memory represents ever existed. I have no senses whatever. Body, shape, extension, movement, and place are all chimeras. What then, will be true? Perhaps just the single fact that nothing is certain (Descartes 1993: AT VII, 24).

This is Cartesian skepticism. It is methodical and it is extreme. Far from the ordinary instances of doubt we may encounter in our everyday lives, this sort of doubt is *extra*-ordinary. It forces one to question whether a person can be certain *anything* exists—including one's self. Yet, despite how far removed we may feel from this form of doubt, Cartesian skepticism *does* apparently yield a secure starting point for certainty. For although it has us question everything, the very fact that we are able to doubt *anything* assumes certainty of our (mental) existence. Doubting, after all, is a form of thinking, so it stands to reason that if I am a doubting thing, I must be a *thinking* thing, and if I am a thinking thing then I have to be certain of my own

[1] Portions of this chapter were initially published in my book, *Exploring Certainty: Wittgenstein and Wide Fields of Thought* (Lexington Books, 2014). My thanks to the editors at Lexington for allowing me to reprint some of that work here.



R. G. Brice, *Wittgenstein's On Certainty: Insight and Method*, SpringerBriefs in Philosophy, https://doi.org/10.1007/978-3-030-90781-5_3

existence (even if that thought is a doubt about my own existence).[2] Hence, the proposition "I think, therefore I am" (*cogito ergo sum*) must be true "every time I utter it, or conceive it in my mind" (Descartes 1993: AT VII, 25). Such a proposition, Descartes reasons, cannot possibly be doubted and therefore must be absolutely certain.

While Descartes' *cogito*—as it has come to be known—seems like a secure foundation for knowledge,[3] paradoxically, it produces some rather skeptical views. G. E. Moore cites "two main varieties:"

> The first variety of this type is that which asserts that we simply do not know at all whether there are any material objects in the Universe at all. It admits that there may be such objects; but it says that none of us knows that there are any…And the second view goes even further than this. It denies also that we can know of the existence of any minds or acts of consciousness except our own. It holds, in fact, that the only substantial kind of thing which any man can know to be in the Universe is simply his own acts of consciousness (Moore 1953: 32).

The first variety is sometimes characterized as "the problem of the external world," for the *cogito* still leaves open the possibility that everything outside of me, i.e., external bodies, including my own body, physical objects, other minds, etc., is illusory. The second variety is known as "the problem of solipsism," and it assumes that, as far as I can know, my mental states are the only mental states that exist. For these and similar reasons, Moore and other academic philosophers in the late nineteenth and early twentieth centuries, objected to Cartesian skepticism. It is "amazing," says Moore, that Descartes, and philosophers in the Cartesian tradition, would

> go beyond or positively contradict the views of Common Sense…[for] they profess to know that there are not in the Universe things of the existence of which Common Sense is most sure (Moore 1953: 14).

Working in the opposite direction, Moore insisted that because propositions of Common Sense are basic, we should not—indeed *cannot*—doubt them. Because Common Sense propositions are fundamentally basic, not only can we *not* doubt them, we cannot offer them any support; we simply know them, they are, as he says, "self-evident"(Moore 1993: 193).

Wittgenstein believed there was something right about using Common Sense to dispute Cartesian skepticism; however, he took issue with Moore's approach, which he found both "unjustified and presumptuous" (Wittgenstein 1972: §553). "Unjustified" because, in the face of this skeptical opposition, Moore failed to

[2] In fact, in the *Third Meditation*, Descartes says, from "the fact that I doubt, it follows that I am" (Descartes 1993: AT VII, 38). Anticipating Descartes' *cogito* argument by one thousand years, St. Augustine wrote, "I am not at all afraid of the arguments of the Academicians, who say, What if you are deceived? For if I am deceived, I am. For he who is not, cannot be deceived; and if I am deceived, by this same token, I am. And since I am if I am deceived, how am I deceived in believing that I am? for it is certain that I am if I am being deceived" (St. Augustine 2017: 316).

[3] Wittgenstein later shows that Descartes' proposition is not as secure as one might think, for Descartes assumes that his language has meaning. But as Wittgenstein says in *On Certainty* §114: "If you are not certain of any fact, you cannot be certain of the meaning of your words either" (Wittgenstein 1972: 114).

provide an adequate explanation of *how* he knew his Common Sense propositions were true. And "presumptuous" because, despite his lack of justification, Moore boldly and repeatedly claimed that he nonetheless *knew* his Common Sense propositions.

## 3.2 Descartes and the Problem of Skepticism

There are, for Descartes, three increasingly skeptical levels of methodological doubt:

(1) *Senses*:

> Surely whatever I had admitted until now as most true I received either from the senses or through the senses. However, I have noticed that the senses are sometimes deceptive; and it is a mark of prudence never to place our complete trust in those who have deceived us even once (Descartes 1993: AT VII, 18).

(2) *Dreams*:

> As I consider these matters more carefully, I see so plainly that there are no definitive signs by which to distinguish being awake from being asleep (Descartes 1993: AT VII, 19).

(3) *Evil genius*:

> I will suppose…an evil genius, supremely powerful and clever, has directed his entire effort at deceiving me…[E]even if it is not within my power to know anything true, it is certainly within my power to take care resolutely to withhold my assent to what is false, lest this deceiver, however powerful, however clever he may be, have any effect on me (Descartes 1993: AT VII, 22–23).

At this point, he says,

> I will…put aside everything that admits of the least of doubt, as if I had discovered it to be completely false. I will stay on this course until I know something certain, or, if nothing else, until I at least know for certain that nothing is certain (Descartes 1993: AT VII, 24).

Here, Descartes presents his most important and captivating argument. If a "supremely powerful and supremely sly" deceiver does indeed exist, and if this deceiver is in fact attempting to deceive Descartes about his own existence, he

> will never bring it about that I am nothing, so long as I shall think that I am something…[Hence,] the pronouncement 'I think, therefore I am' is necessarily true every time I utter it, or conceive it in my mind (Descartes 1993: AT VII, 25).

Let's consider what this means. Descartes insists he is certain of his own "existence," however, he is not referring to physical existence, he admits he can be certain only that he exists as a *thinking being*.

> At this time, I admit nothing that is not necessarily true. I am therefore precisely nothing but a thinking thing; that is, a mind, or intellect, or understanding, or reason…I am a true thing and am truly existing; but what kind of thing?...a *thinking thing* (Descartes 1993: AT VII, 27).

Although Descartes seems to have successfully used his methodological doubt to establish his certainty that he exists as a thing that thinks, this does not mean that all remaining doubt is to be jettisoned. He still must prove that there is a reality external to his mind; failure to do this elicits the problem of the external world (one of the "two main varieties" of skepticism Moore cited above). If Descartes wants to establish that there is an external world, he is going to have to do it through the contents of his own consciousness. He must therefore find something *within* his consciousness that leads him *outside* it. This something, according to Descartes, is God.[4]

Within his mind, Descartes says he has both ideas of perfection and ideas of all the flawless qualities of God. At the same time, he believes that, as a limited and imperfect human being, he does not himself possess these flawless, God-like qualities.[5] Where then could such ideas have come from? These ideas must have come from somewhere else, he reasons. There must be a *cause* (external) for these ideas. "For whence, I ask, could an effect get its reality, if not from its cause? And how could the cause give that reality to the effect, *unless it also possessed that reality*?" (Descartes 1993: AT VII, 40). This Causal Argument for the existence of God,[6] as it has come to be known,[7] holds that what is in the "idea" is contained in whatever brought it about. So, according to Descartes, since he is a finite being, and since he

[4] Having been educated by the Jesuits at La Fleche, this would not be such an odd direction for Descartes to turn, but it may also have been calculated. Descartes wanted to make sure his philosophy was being discussed among *intelligentsia*. As he says at the end of the *Discourse*, Part VI: "I would be very happy if people examined my writings and, so that they might have more of an opportunity to do this, I ask all who have objections to make to take the trouble and send them to my publisher…I shall try to publish my reply at the same time as the objections; by this means, seeing both of them together, the readers will more easily judge the truth of the matter" (Descartes 1998: AT I, 455–456). Why would he do this unless he hoped, as Roger Ariew says, "to promote discussions of his views"? Roger Ariew, "Descartes and Scholasticism: The Intellectual Background to Descartes' Thought," p. 62. Having been educated at Le Fleche, Descartes knew how to talk to the Jesuits, even if what he was saying the Jesuits might see as a danger to their faith. For instance, Descartes rejected a lot of Aristotelian dogma, e.g., his four causes; the four elements: earth, air, fire, and water; he rejected the three Aristotelian principles of *matter, form,* and *privation*. Except for rational beings who have minds, Descartes rejected the doctrine of substantial forms. Finally, though Descartes might've agreed that fire is hot and dry, and air is humid and hot, it would have been as phenomenological descriptions, and not as representing any basic reality; such statements would've been inconsistent with Descartes' mechanical philosophy" (Ariew, 1992: 65–66). These, however, were ideas and concepts Jesuits accepted without question.

[5] While it is possible that thoughts about external objects e.g., the sky, the earth, light, etc. are all delusions of the mind, the same is not possible of God. Why not? Because, he tells us, these other thoughts are of imperfect objects, they could easily be invented by an imperfect mind such as his.

[6] This is also found in *Discourse on the Method*: "I decided to search for the source from which I have learned to think of something more perfect than I was, and I plainly knew that this had to be from some nature that was in fact more perfect…It thus remained that this idea had been placed in me by a nature truly more perfect than I was and that it even had within itself all the perfections of which I could have any idea, that is to say, to explain myself in a single word, that it was God" (Descartes 1998: AT, 33–34).

[7] It should be noted that this argument has the character of an ontological argument, since it starts from the idea of God, and yet it adds the notion of causality (ideas have causes; I have an idea of God; God must exist as the necessary condition for the idea that I possess).

could not have the idea of an infinite being unless this idea proceeded from some being which really was infinite, it follows that this infinite being necessarily exists.[8]

While most of the philosophical community has accepted the formulation of its problems from Descartes, few have accepted his solutions—especially his argument for the existence of God.[9] It is worth pointing out that the proof of God's existence is not a small, self-contained problem in the *Meditations*, it is fundamental to Descartes' entire philosophy. For without an effective proof of the existence of a perfect, and thus non-deceiving God, Descartes believes he will not be able to escape from the corner into which he has painted himself through this radical (or extreme) methodological doubt. And if he is unable to get beyond the foundation of the *cogito*, then he is unable to offer any satisfactory answers to the skeptical questions his method raises, namely,

1. is there a reality external to our minds?

---

[8] More formally, the argument looks something like this:

1. I have an idea of a perfect being (God).
2. In every cause there must be at least as much reality as there is in the effect.
3. If every cause must be at least as great as its effect (from 2), then whatever caused my idea of perfection (from 1) must be perfect.

---

4. ∴ A perfect being (God) exists.

Here's another way to put it: I have this idea of God in my mind, but I also see an absolutely necessary intuitive principle which is that the lesser cannot give rise to the greater, that is, the lesser cannot be the cause of the greater. Now his idea of God is an infinite thing, and although it is only an idea in itself, it is nonetheless, an idea *of* an infinite being. It involves the idea that I can *conceive* an infinite being. But Descartes reasons that no finite being (as he himself is) can give rise to this idea of an infinite being all by himself; it could only have been implanted in him by God himself. When he reflects that the lesser cannot give rise to the greater, Descartes realizes that since he has this idea of God, it can only be because there actually *is* a God who has created him. With this conclusion, Descartes believes he has founded our knowledge of the external world on the "self-evidentness" of the existence of God.

[9] Perhaps the strongest argument *against* the causal argument is one that Descartes should have been familiar with. It was already centuries old by the time he wrote the *Meditations*. It comes from Gaunilo, an eleventh century Benedictine monk, best known for his criticism of Anselm's argument for the existence of God, an argument that is quite similar to Descartes' causal argument. Gaunilo's counter-argument shows the falsity of Descartes' third premise. I may have the concept of a perfect island, argues Gaunilo, but this concept (effect) does not mean a perfect island actually *exists* (cause). But according to Descartes' reasoning, the perfection of this island would therefore *imply* that it *exists*. Possessing a concept of something, however, does not make what this concept refers to *exist*, nor does it make it exist by simply adding the attribute of perfection to it. Notice, the causal argument suggests a strong link between the cause of an object and its effect. But here's another problem. Consider some of the following counter-examples to this: While a bridge, in-and-of-itself may be strong, the ingredients of a strong bridge do not *themselves* contain strength. Sponge cake has many properties *not* present in the ingredients (e.g. sponginess). Finally, there is David Hume who argued that the idea of God could be arrived at by considering qualities within oneself (wisdom, strength, goodness) and magnifying them. Knowledge of God, therefore, is not innate, rather, it is taught to us. All ideas, says Hume, come from impressions.

And if so,

2. can we have knowledge of this reality?

Imagine, for a moment, that the certainty of the contents of your own mind is the total and complete extent of your knowledge. What might this be like? Martin Benjamin once characterized Descartes' precarious position here as that of a "lone, disembodied spectator."

> *Alone* because, in doubting whatever can possibly be doubted, you're doubting that other people exist...*disembodied* [because] you can be mistaken about the reality of your own body for the same reason you can be mistaken about the existence of other people (and their bodies)...[And] a *spectator* because without a body-and without other physical objects in the 'external' world-there is nothing, except reflecting on one's own thoughts and beliefs, one can do (Benjamin 2003: 17).

Let's now turn to Moore and his Common Sense reply to Descartes' sweeping form of skepticism.

## 3.3 Moore's Common Sense Reply to Cartesian Skepticism

Moore was not the first to advocate Common Sense in an effort to combat radical skepticism. As we've seen, this thread can be traced back to the eighteenth century and Thomas Reid's Scottish School of Common Sense.[10] Against such skepticism, recall, Reid had claimed

> [t]he evidence of sense, the evidence of memory, and the evidence of the necessary relations of things, are all distinct and original kinds of evidence, equally grounded on our constitution...To reason against any of these kinds of evidence is absurd...They are first principles; and such fall not within the province of reason, but of common sense (Reid 2012: 44–45).

Similarly, by Common Sense, Moore is referring to those "things which we all commonly assume to be true about the Universe, and which we are sure that we know to be true about it" (Moore 1953: 14). For example, we are certain of the existence of

> our own bodies, the bodies of millions of others, the bodies of millions of animals, of plants, of inanimate objects, e.g., mountains, stones, sand, minerals, soils, water in rivers and seas (Moore 1953: 15).

We know that there is an earth and other heavenly bodies, e.g., the sun, the moon, the visible stars, etc. We know that there are many different objects that we humans have made, including—but obviously not limited to—houses, chairs, tables, railway engines (Moore 1953: 15). These propositions are obvious truisms; indeed, they are so obvious that we hardly find occasion to utter them.[11]

---

[10] See Chapter 2.2.

[11] Moore sometimes provides a list of common sense propositions that he knows to be true. For instance, in "Defense" he says, "There exists at present a living human body, which is *my* body,... there [are] large numbers of other living human bodies, each of which has...at some time been

Moore says that because a Common Sense proposition is basic, we cannot offer any further support; we simply know it. Take our conviction that external facts—facts outside of one's own mental state—exist. "The only proof that we do know external facts," says Moore "lies in the simple fact that we do know them" (Moore 1909: 160). In this way, we can offer no further support; Common Sense propositions, he says, are "self-evident." By "self-evident," he means that

> the proposition so-called is evident or true by itself alone; it is not an inference from some proposition other than itself…By saying that a proposition is self-evident, we mean emphatically that its appearing so to us is not the reason why it is true: for we mean that it has absolutely no reason (Moore 1993: 193).

So, while self-evident propositions are true, our reason—our justification for holding them true—is that there is no reason. This is the point at which Moore terminates the investigation. As he says in another work, "I do not know exactly how to set about arguing that [Common Sense propositions] are self-evident" (Moore 1919: 191). In his two most well known articles, "Proof of An External World" and "A Defense of Common Sense," he repeatedly makes this very point. In "Proof" he says:

> How am I to prove now that 'Here is one hand, and here's another'? I do not believe I can do it…I could not tell you what my evidence is; and I should require to do this at least, in order to give you a proof…[However,] I can know things which I cannot prove; and among things which I certainly did know, even if (as I think) I could not prove them, were the premises of my…proof (Moore 1939: 148).

And in "Defense" he says:

> We are all, I think, in this strange position that we do know many things, with regard to which we know further that we must have had evidence for them, and yet we do not know how we know them; i.e., we do not know what the evidence was (Moore 1925: 44).

We might be tempted to think that Moore is more accurately expressing his *belief* of Common Sense propositions, not his knowledge of them. Yet, he categorically denies this:

> Some…have spoken of such beliefs as "beliefs of Common Sense," expressing thereby their conviction that beliefs of this kind are very commonly entertained by mankind: but they are convinced that these things are, in all cases, only believed, not known for certain; and some have expressed this by saying that they are matters of Faith, not Knowledge (Moore 1925: 42).

---

born,...continued to exist for some time after birth,...been, at every moment of its life after birth, either in contact with or not far from the surface of the earth; and many of these bodies have already died and ceased to exist…the earth ha[s] existed…for many years before my body was born…large numbers of human bodies ha[ve]…been alive upon it; and many of these bodies hav[e] died and ceased to exist before [my body] was born…I am a human being, and I have…had many different experiences,…e.g., I have often perceived both my own body and other things which formed part of its environment, including other human bodies; I have…also observed facts about them…And, just as my body has…had many experiences of each of these (and other) different kinds; so, in the case of very many of the other human bodies which have lived upon the earth, each has…had many different experiences of each of these (and other) different kinds" (Moore, 1925: 33–34).

The problem, says Moore, is that skeptics "deny that we ever know [Common Sense propositions], for certain, to be true." And to this, he says, "I have nothing better to say than that it seems to me that I do know them with certainty" (Moore 1925: 41).

But there is a problem here. The guarantee that he knows something—that he has two hands, for example—does not ensure that Moore really *does* know it, only that he *thinks* he knows it. "That he *does* know takes some showing," says Wittgenstein (1972: §14, my emphasis). And showing means that Moore must go further than just insisting that he is right.

## 3.4 Problems in Moore's "Proof of an External World"

There are other related problems that emerge in Moore's writing—specifically in "Proof of An External World." In this article, Moore sought a proof to meet Kant's challenge for "the existence of things outside of us"—external to the mind. Quoting Kant, Moore says:

> It still remains a scandal to philosophy…that the existence of things outside of us…must be accepted merely on faith, and that, if anyone thinks good to doubt their existence, we are unable to counter his doubts by any satisfactory proof (Kant 1965: 34).

This, of course, is Kant's version of the Cartesian problem of the external world. But Moore believed he could remove this "scandal," indeed solve the problem, with proof of the existence of his own two hands:

> I can prove now…that two human hands exist. How? By holding up my hands and saying, as I make a certain gesture with the right hand, "Here is one hand," and adding, as I make a certain gesture with the left, "and here is another"…[This] proof…was a perfectly rigorous one; and…it is perhaps impossible to give a better or more rigorous proof of anything whatever (Moore 1939: 144).

As it stands, Moore's argument can be reduced to the following basic structure:

1. Here is one hand.
2. Here is another (hand).

---

3. ∴ Two hands exist.[12]

---

[12]One quick point of clarification. Moore seems to have interpreted the Kantian phrase "x is outside of us" to mean "x is an external object." Avrum Stroll offers this clarification (Stroll 1994: 56).

Although this is a valid *argument*,[13] the concern is whether it can establish that something exists outside the mind. Can we arrive at the conclusion that "two hands *exist*" from our premises? Has Moore established that a world exists independent of our minds? More to the point: has Moore offered a *proof*?

Moore's demonstration suffers from two large problems. First, each premise in this argument already *assumes* that something exists—one hand (and another hand). But this just begs the question he sought to answer, namely, does something exist *outside* the mind? Second, if we were to charitably remove the metaphysical assumptions here in the premises (the assumptions that create the question-begging), it appears that Moore's "proof" is nothing more than the simple mathematical application of 1 + 1 = 2. But simple arithmetic does not *prove* that things *exist* "outside of us." That is to say, mathematics alone does not necessarily allow us to establish any conclusions about the nature or existence of the external world. How then do we *know* Moore's premises are true?

In order for us to know Moore's premises are true, he must present reasons (evidence) why we should accept his premises. But Moore refuses to do this, in fact, at the end of "Proof," he confesses that he cannot:

> How am I to prove now that 'Here is one hand, and here's another'? I do not believe I can do it...I could not tell you what my evidence is; and I should require to do this at least, in order to give you a proof (Moore 1939: 148).

Yet, even with this admission, Moore goes on to make this astonishing statement: "I can know things which I cannot prove; and among things which I certainly did know, even if (as I think) I could not prove them, were the premises of my…proof" (Moore 1939: 148).[14] This argument in defense of his "proof" is unsatisfactory, for Moore makes assumptions where he *should* offer definitive reasons. Thus, his proof fails.

It appears Moore was well aware philosophers would take issue with his proof. As he says, what they really want

> is not merely a proof…but something like a general statement as to how any propositions can be proved. This, of course, I haven't given; and I do not believe it can be given: if this

[13] There are, of course, some assumptions in this argument that Avrum Stroll patiently brings into the light. He refers to these as "submerged premises"; these include: (1) The existence of any hand is not mind-dependent. (2) Anything whose existence is not mind-dependent exists outside of us. Moore, says Stroll, "imported these without acknowledgment into his…argument, so that, in effect, they function as enthymemes" (Stroll 1994: 56). If we were to charitably couple these submerged premises with Moore's original premises, the argument is unmistakably valid: (3) Here is one hand. (4) Here is another (hand). (5) ∴ Two hands exist outside of us.

[14] But this argument begs our question for *proof* of an external world. What is perhaps most devastating here for Moore is that by assuming this split between an internal and external world, he commits himself to the "intellectualized" tradition he so disliked. Implicit in Kant's challenge was a distinction between what is internal and external to our minds. But in offering a proof for "things outside us," Moore tacitly accepts this division. As Stroll says, "Moore is working with two metaphysical categories, the mental and the physical, and is trying to draw an exact line between them."

> is what they mean by proof of the existence of external things, I do not believe that any proof of the existence of external things is possible…[Yet] I can know things which I cannot prove (Moore 1939: 147).

"A general statement" is reminiscent of Descartes who, at the beginning of the *Second Meditation*, offered a general criterion of knowledge, one in which he declared that certain clear and distinct ideas were true.[15] This, Moore refused to do.[16] For had he *generally stated* that he trusts his senses (trusts that he *sees* his hands, for example), the skeptic would undoubtedly have asked Moore whether his senses had ever deceived him. Other aspects of Cartesian skepticism would then emerge, as he says,

> I should need to prove…as Descartes pointed out, that I am not now dreaming. But how can I prove that I am not? I have, no doubt, conclusive reasons for asserting that I am not now dreaming; I have conclusive evidence that I am awake: but that is a very different thing from being able to prove it (Moore 1939: 148).

Moore's "conclusive evidence" would simply set up the skeptic for another round of inquiry. But regardless of the evidence he might put forward, the skeptic would always have the last word. In order to avoid this, Moore instead declared that he could not be wrong about his premises and, in turn, that he could not be wrong about his proof.

But a proof lacking sufficient support falls short of meeting Kant's challenge.[17] "I know that *p*" cannot be the epistemological end-point of an investigation; it does not satisfy skeptical curiosity, in fact, it has the opposite effect. Yet, with "I know that I have two hands," Moore seemed to be offering a *guarantee* that what he "knew" was an unquestionable, indubitable fact.[18] But the question is: why should a skeptic believe him? Why should Moore's assurance that he "knows" this or that be a reliable source of justification for us? Affirmation and reiteration will not help. In order for him to succeed he would have to be able to prove *how* he knows what he *says* he knows; otherwise accepting Moore's assurances would be no different than

---

[15] "I am certain that I am a thing which thinks; but do I not then likewise know what is requisite to render me certain of a truth? Certainly in this first knowledge there is nothing that assures me of its truth, excepting the clear and distinct perception of that which I state, which would not indeed suffice to assure me that what I say is true, if it could ever happen that a thing which I conceived so clearly and distinctly could be false; and accordingly it seems to me that already I can establish as a general rule that all things that I perceive very clearly and distinctly are true" (Descartes 1993: AT VII, 35).

[16] Aware of the viciously circular argument a general statement created for Descartes (also known as the "Cartesian Circle"), Moore refused to offer one.

[17] Some of his contemporaries said as much. Max Black went even further, condescendingly describing Moore's argument as childlike: "After the intoxication of metaphysics," Black wrote, "it is good to look upon the world again as a child might-to be told 'After all, this *is* a hand. I have a body, so have you, and there are many other people like both of us who can say the same.'" Black, Max. 1971. *Philosophical Analysis: A Collection of Essays*. Books For Libraries, p. 7.

[18] Although we may be tempted to think that knowledge *guarantees* what is known, guarantees it as a fact, Wittgenstein reminds us, we "always forget the expression 'I *thought* I knew'" (Wittgenstein 1972: §12, my emphasis).

accepting what he says "on faith." And here we return to an issue that often emerges in Moore's articles—lack of justification for his Common Sense propositions (as we saw in Sect. 3.3, above).[19]

Early in *On Certainty*, with Moore no doubt in mind, Wittgenstein makes the following observation: "It would surely be remarkable if we had to believe the reliable person who says 'I can't be wrong'; or who says 'I am not wrong'" (Wittgenstein 1972: §22).

## 3.5 Wittgenstein's Therapeutic Conception of Philosophy

Wittgenstein agreed with much of the Common Sense approach and thought it might even be possible to "defend common sense against the attacks of the philosophers," but he insisted this could only be done by first "curing them of the temptation to attack common sense, not by restating the views of common sense" (Wittgenstein 1958: 58–59). Just stating, or *re*-stating Common Sense truisms is not enough. We must look at the *cause* of this temptation, a temptation he characterizes as a kind of "sickness" or "illness." It is possible, he says, "for the sickness of philosophical problems to get cured [but] only through a changed thought and life" (Wittgenstein 1998: II, §23). For Wittgenstein, an adjustment such as this meant a shift not only in how we typically confront radical skepticism and how the problem is resolved/dissolved, but also a close examination of the discipline of philosophy.

Philosophers who comprehensively question our Common Sense convictions about the world—those caught in the grip of Cartesian skepticism—are not merely committing reasonable "mistakes"; they are suffering from, what Wittgenstein variously describes as, a "disease of thought" (1967: §382) a "disease of the understanding" (1978: §50), or sometimes, a "philosophical disease" (1953: §593). It is hard to take the Cartesian skeptic seriously when he questions such basic convictions, like the earth's existence. Can we doubt the earth's existence while we simultaneously *lead our lives*? Academic philosophers may enjoy pushing the boundaries of what is logically permissible, but Wittgenstein believes this has been permitted for too long. In any *other* discipline, we might conclude that this person is suffering from some kind of "mental disturbance" (1972: §71). In order for this sickness to be treated, Wittgenstein recommends a kind of *therapeutic philosophy*—one that will help "cure [the] many diseases of the understanding," and bring us back to "the notions of common sense" (1984: §50).

The kind of therapeutic philosophy he has in mind will force us to change how we think about (and attempt to resolve) traditional philosophical issues like radical

[19] It occurs in "Defense," "[w]e are all…in this strange position that we do *know* many things…and yet we do not know *how* we know them" (Moore, 1925: 44). And in "Proof," "How am I to prove now that 'Here is one hand, and here's another'? I do not believe I can do it...I could not tell you what my evidence is; and I should require to do this at least, in order to give you a proof…I can know things which I cannot prove" Moore (1939: 148).

Cartesian skepticism; it will also force us to examine how our philosophical convictions affect (or fail to affect) our daily lives. In this way, we might think of Wittgenstein's therapeutic method as consisting in two broad goals:

(1) to bring about an awareness in the philosopher that Cartesian skepticism is not a methodological approach to be taken seriously and, with this awareness,

(2) an attempt to replace this radical skepticism with a practical Common Sense groundwork.

Like clinical, psychological therapy, *philosophical* therapy consists in fostering a new way of looking at things, a corresponding new way of thinking, and, in turn, a new way of life.

These two approaches—psychological and philosophical therapy—closely resemble one another.[20] For the remainder of this chapter, I offer a brief comparison of the aims found in the psychotherapy introduced in Chap. 2, *Cognitive Behavioral Therapy*, with those found in Wittgenstein's therapeutic method.

Let's begin with the National Institute of Mental Health's (NIMH) definition of "psychotherapy":

> Psychotherapy is a way to treat people with a mental disorder by helping them understand their illness. It teaches people strategies and gives them tools to deal with stress and unhealthy thoughts and behaviors. Psychotherapy helps patients manage their symptoms better and function at their best in everyday life.[21]

In *Cognitive Behavioral Therapy* (or CBT), the patient and therapist, together, walk through four steps:

1. Identify troubling situations or conditions in your life.
2. Become aware of your thoughts, emotions and beliefs about these problems.
3. Identify negative or inaccurate thinking.
4. Reshape negative or inaccurate thinking.[22]

One of the first things the therapist must do is unearth what is hidden beneath the surface of the patient's beliefs and behaviors before carefully and cautiously trying to adjust them. Once the problem has been identified, the therapist attempts to alter

[20] In multiple places in his writing, Wittgenstein compares his philosophical method to a kind of therapy (or therapies). See *Philosophical Investigations* §133, §§254–255, *Zettel* §382 *Big Typescript* 407–410. More often than not, however, scholars tend to focus on the similarities between his method and that of Freudian psychoanalysis, see work by Cora Diamond, James Conant, and Stanley Cavell. While these authors shed a significant amount of light on the similarities here, my analysis seeks to compare Wittgenstein's method to another form of therapy: *Cognitive Behavioral Therapy* (or *CBT*). As Wittgenstein says in *Philosophical Investigations* §133: "There is not *a* philosophical method, though there are indeed methods, like *different therapies*" (second emphasis is mine).

[21] This definition is located on the National Institute of Mental Health's website: www.nimh.nih.gov/health/topics/psychotherapies/index.shtml.

[22] These steps, as well as what one can expect at each, can be found on the Mayo Clinic's website: https://www.mayoclinic.org/tests-procedures/cognitive-behavioral-therapy/about/pac-20384610

what the patient takes to be normal functioning and reshape it into something new. Reshaping stultifying patterns of thinking is a difficult step. The patient has spent his/her life thinking about this issue one way. But with practice, new behavioral patterns will become habit.

When successful, this new functioning ultimately *becomes* normal; the result will establish a deep-seated change in the person being treated. In this way, therapy effects a change in the *sensibility* of the patient. Perhaps most importantly, however, is the fact that through a discursive process between patient and therapist the patient comes to recognize *what* his (hidden) patterns of beliefs are, *why* they are destructive, and *how* (with the help of the therapist) he can develop tools to manage and function whenever these beliefs and behaviors re-emerge.

In his book, *Ethics Without Philosophy*, philosopher James Edwards discusses what the therapist hopes to accomplish when faced with a particular kind of pathology: compulsive hand-washing. Attempting to convince the patient that his hands are clean will prove to be fruitless.

> The object of the therapist [here] cannot felicitously be described as returning the patient to his "normal functioning." In a very real sense, for this patient, compulsive hand-washing is his "normal functioning." For a person of his sensibilities and with his past, some pathological behavior is inevitable. What is needed now, of course, is not just the extinction of the pathological behavior,…what is needed is a change in what counts as the patient's "normal functioning," so that the pathological symptoms will truly disappear and thus will not return in another form. Since life's inevitable traumas cannot be removed, what is needed is a change in (what might be called) the sensibility of the patient. He needs to be equipped to weather these traumas with limited impairment, and this can only be accomplished if he becomes a different sort of person (Edwards 1985: 134).

Like the therapist to his patient, Wittgenstein's objective cannot be to have the philosopher return to his "normal functioning." His goal cannot be to have the philosopher exceed the boundaries of Common Sense and return to posing impractical (or dead-end) questions—all within this unworkable (Cartesian) system. This may be where he comfortably "functions," that is, philosophizing within this system may be where the "sick" philosopher feels most at ease, but what he produces here has proved to be ineffective and costly, overall, to the discipline by making it look merely "academic," in a pejorative sense of the word. If philosophy is ever to be worthwhile and productive once again, if it is to be more than thought experiments and mental exercises, it must work to serve *practical ends*. This, however, first requires a "change in the sensibility of the patient."

The goal of therapeutic philosophy then is to change the philosopher's self-defeating patterns of thinking, thus changing his behavior. What is needed is not just his recognition that this particular belief is unhealthy, but a recognition that his *pattern of thinking* is unhealthy—a *recognition* that his thinking is impractical, in the sense of having little or nothing to do with how we live our lives. If the therapist can succeed in this goal, this may serve as a catalyst for change in the philosopher's behavior. This change in behavior requires a willingness to *reshape* the old way of thinking into something that is not only healthy, new, and practical, but something that will serve to *become* his "new normal."[23]

Philosopher Eugen Fischer has said the problems that emerge when we reflect philosophically within the Cartesian framework are not problems at all, they are "pseudo-problems" or, what he calls, "philosophical monsters."

> Where we are worried by a real problem, we need to solve the problem. But where we are worried by an imaginary problem we have no right to believe confronts us, the only thing we can reasonably do is to get ourselves to stop unreasonably worrying about it. (Compare: When a child is afraid that monsters might creep in at night, there is no real security problem; we do not have to guard the room, we have to liberate the child from his fear.) The philosophical problems that Wittgenstein has in mind are of this kind: We formulate them, he thinks, as a result of certain misunderstandings, which have us imagine problems where there are none ("philosophical monsters," as it were). When such a pseudo-problem worries us, the only problem we actually face is the emotional problem that consists in the unwarranted feelings of disquiet, which our misunderstandings…cause in us (Fischer 2011: 58).

Psychotherapy seeks to rid us of our "monsters" by readjusting our beliefs and behaviors in deep and significant ways; ways that get us to see that these monsters do not exist.

One of the most important steps in this therapeutic process is getting us to comprehend what is wrong—"what is it I am suffering from?" Answering this question takes a lot of work and a lot of time. Therapy is a slow process in which the patient is brought—by degrees—to a new understanding of the nature of the problem/s that trouble him.[24] Similarly, when "philosophizing," says Wittgenstein, "we may not terminate a disease of thought. It must run its natural course, and slow cure is all important" (1967: §382). This "slow cure" requires patiently and repeatedly returning to the "diseased thought" with our new tools that serve to demonstrate why this pattern of thinking is harmful.

The tools that will lead to this new "normal functioning" in the philosophical patient include: a recognition on his part that there is a distinction between ordinary skeptical questions (what Wittgenstein calls "mistakes") and extraordinary ones ("mental disturbances"), and a recognition that the latter kind of doubt tends to lead one down the path to unhealthy, unfruitful (Cartesian) thought and behavior. When this happens, the philosophical patient must be aware of the danger of heading down this path. *Awareness* that this is occurring is the first and most important tool. One must be capable of seeing this and be capable of adjusting one's behavior.[25]

[23] This speaks directly to CBT's goal. Again, see the Mayo Clinic's website: https://www.mayoclinic.org/tests-procedures/cognitive-behavioral-therapy/about/pac-20384610

[24] An advantage of CBT over the older Freudian psychotherapy seems to be its efficiency and practicality.

[25] In this way, the change is more accurately characterized by the more practical conception of therapy as cognitive (CBT), than by the often questionable conception of Freudian psychotherapy (involving the unconscious, childhood experiences, etc.). My thanks to Martin Benjamin for pointing this out to me.

In the following chapter, I discuss one of the major themes found in *On Certainty*: mistake vs. "mental disturbance." This distinction helps illustrate the need for a therapeutic approach to certain philosophical questions.

## References


Ariew, Roger. 1992. Descartes and Scholasticism: The Intellectual Background to Descartes' Thought. In *Cambridge Companion to Descartes*, ed. J. Cottingham. Cambridge: Cambridge University Press.

Augustine of Hippo, St. 2017. (Originally published 1470). *The City of God*, trans. Marcus Dods. Digireads.com Publishing.

Benjamin, Martin. 2003. *Philosophy & This Actual World*. Lanham: Rowman & Littlefield Press.

Descartes, Rene. 1993. (Originally published 1641). *Meditations on First Philosophy*, 3rd ed, trans. Donald Cress. Indianapolis: Hackett Publishing.

———. 1998. (Originally published 1637). *Discourse on the Method*, 3rd ed, trans. Donald Cress. Indianapolis: Hackett Publishing Company.

Edwards, James. 1985. *Ethics Without Philosophy: Wittgenstein and the Moral Life*. Miami: University Press of Florida.

Fischer, Eugen. 2011. How to Practice Philosophy as Therapy: Philosophical Therapy and Therapeutic Philosophy. *Metaphilosophy* 42 (1–2): 49–82.

Kant, Immanuel. 1965. (Originally published 1781). *Critique of Pure Reason*, trans. Norman Kemp Smith. New York: St Martin's Press.

Moore, G.E. 1909. Hume's Philosophy. In *Philosophical Studies*. London: George Allen and Unwin Ltd.

———. 1919. External and Internal Relations. In *Philosophical Studies*. London: George Allen and Unwin Ltd.

———. 1925. Defense of Common Sense. In *Philosophical Papers 1962*. New York: Collier Books.

———. 1939. Proof Of An External World. In *Philosophical Papers 1962*. New York: Collier Books.

———. 1953. What is Philosophy? In *Some Main Problems of Philosophy*. New York: Collier Books.

———. 1993. (Originally published 1903). *Principia Ethica*. Cambridge: Cambridge University Press.

Reid, Thomas. 2012. *Selections from the Scottish Philosophy of Common Sense*. Chicago: Veritatis Splendor Publications.

Stroll, Avrum. 1994. *Moore and Wittgenstein on Certainty*. Oxford: Oxford University Press.

Wittgenstein, Ludwig. 1953. *Philosophical Investigations*, trans. G.E.M. Anscombe. Oxford: Blackwell Publishers.

———. 1958. *The Blue and Brown Books. Preliminary Studies for the Philosophical Investigations*. Oxford: Basil Blackwell.

———. 1967. *Zettel*, 2nd ed, eds. G.E.M. Anscombe and G.H. von Wright, trans. G.E.M. Anscombe. Oxford: Blackwell Publishers Ltd.

———. 1972. *On Certainty*, eds. G.E.M. Anscombe and G.H. Von Wright, trans. D. Paul and G.E.M. Anscombe. New York: Harper & Row.

———. 1978. *Philosophical Grammar*, ed. R. Rhees, trans. Anthony Kenny. Los Angeles: University of California Press.

———. 1984. *Culture and Value*, eds. G.E.M. Anscombe and G.H. von Wright, trans. Peter Winch. Chicago: University of Chicago Press.

———. 1998. *Remarks on the Foundations of Mathematics*, eds. G.E.M. Anscombe, G.H. Von Wright, and Rush Rhees. Cambridge: MIT Press.

# Chapter 4
# Philosophical Therapy: A Cure for Our Philosophical Disease

**Abstract** This chapter first offers an account of Wittgenstein's general therapeutic approach as it applies to knowledge and the skeptical challenge before turning to its application in *On Certainty*.

**Keywords** Cognitive Behavioral Therapy (CBT) · Common Sense · Philosophical disease · G.E. Moore · Knowledge · Mistake · Mental disturbance · Reshape · Skepticism · Troubling · Ludwig Wittgenstein

## 4.1 Mistake Versus "Mental Disturbance"

Imagine staring at your hand while uttering the statement, "I know that here is a hand." Now imagine someone *doubting* that you knew such a thing, "But do you *really* know that? Couldn't you be wrong?" What would you *say* to this person? Maybe something like, but "it's *my* hand I'm looking at!" (Wittgenstein 1972: §19). Whatever your reply, you would probably question his sanity. Is he rational? Is he reasonable? For surely, the "reasonable man," as Wittgenstein says, "will not doubt that I know" (Wittgenstein 1972: §19). Unfortunately, for centuries now, philosophy has given equal footing to both reasonable and unreasonable doubt. Wittgenstein's philosophically adroit sense of humor captures this nicely:

> I am sitting with a philosopher in the garden; he says again and again "I know that that's a tree," pointing to a tree that is near us. Someone else arrives and hears this, and I tell him: "This fellow isn't insane. We are only doing philosophy" (Wittgenstein 1972: §467).

In several passages in *On Certainty*, Wittgenstein discusses why unreasonable skepticism is problematic; however, there are a handful of passages (§§71–74 and §§155–157 in particular), where he draws a key distinction between a *mistake* and what he calls a *mental disturbance*. This distinction will serve as a point of entry for us as we begin thinking about legitimate forms of philosophical inquiry and the


R. G. Brice, *Wittgenstein's On Certainty: Insight and Method*, SpringerBriefs in Philosophy, https://doi.org/10.1007/978-3-030-90781-5_4

illegitimate sort that permit philosophical diseases like radical skepticism to take root and spread.

We can be mistaken about the dates of an ancient war, the results of a chemistry experiment, or about whether a tree we happen to be looking at is a beech or an elm. To be mistaken about these sorts of things may elicit a kind of doubt in us the next time we are asked, say, when the Peloponnesian war took place, what occurs when one mixes a small amount of sodium chlorate and sugar, with a few drops of sulfuric acid, or—when confronted with a tree—whether it is a beech or an elm. But these kinds of doubt are just the sort we might expect; they are typical, ordinary. Now contrast a mistake with an error so abnormal that it threatens our basic, Common Sense convictions; an error so extraordinary the person committing it seems to have lost contact with the world. This person is not committing a mistake, according to Wittgenstein he is instead suffering from a kind of "mental disturbance."

> If my friend were to imagine one day that he had been living for a long time past in such and such a place, etc. etc., I should not call this a mistake, but rather a mental disturbance (Wittgenstein 1972: §71).

Imagine how impatient we'd become with someone who questioned whether he'd lived in London for the last 20 years or in Calcutta. When we encounter a person like this, we tend to respond as Wittgenstein does in *Last Writings On The Philosophy Of Psychology*,

> half annoyed and half embarrassed...All the while we [are] conscious that on the one hand we are not at all capable of giving reasons...because seemingly there are too many, and on the other hand that no doubt is possible (Wittgenstein 1992: §53).

While Wittgenstein's use of the phrase "mental disturbance" has the potential to be both offensive and stigmatizing, it should not be interpreted this way. By "mental disturbance," he means to highlight a doubt so peculiar that we might characterize it, like Avrum Stroll does, as something "aberrant, strange, incomprehensible" (Stroll 1994: 47).[1] The difference between a mistake and a mental disturbance is perhaps most helpful in what it reveals about philosophy, specifically, the extraordinary levels philosophers are willing to push inquiry. For the radical skeptic—the one who doubts *everything*—to even *pose* his question demands he first accept *something* about the world, namely, that there *is* one. If we were unable to get him to see this, if he were to seriously entertain such a doubt, we would not think him wrong, but suffering from a "philosophical disease" (Wittgenstein 1953: §593).

While it is one thing to *identify* a disease, it is quite another to *treat* it. Wittgenstein does more than simply identify the problem, his prescription, what I am calling *therapeutic philosophy*,[2] requires we no longer *engage* the philosopher in serious philosophical discourse; instead, we must treat him like a person afflicted with a

[1] Furthermore, many people would call affective disorders (like depression) mental disturbances, yet I do not want to imply that people who are depressed are irrational to the extent that Wittgenstein is using the phrase.

[2] I will use the terms "therapeutic philosophy" and "philosophical therapy" interchangeably throughout.

disorder, a disorder that has proven detrimental to his intellectual well-being. To attend this sickness, Wittgenstein recommends a slow but steady *re*adjustment, a shift in how he confronts philosophical problems. This readjustment comes in a series of stages that, if successful, will ultimately serve to reshape the negative or inaccurate thinking.

As mentioned in Chap. 3, it is difficult to take the Cartesian skeptic seriously when he questions our most basic convictions. And although the skeptic may be engaging in philosophical speculation that conforms to traditional rules of the discipline—i.e., placing equal weight on all logical possibilities (anything that might, however probable or bizarre, possibly be true)—he remains stuck in an isolated, solipsistic world. This world is one where even our basic, Common Sense convictions are targeted and questioned. As it turns out, this is symptomatic of the philosopher's "disease." What is needed is a therapeutic approach to resulting philosophical problems, one that will, as Wittgenstein says, "cure [the] many diseases of the understanding," one that will return us to "the notions of common sense" (Wittgenstein 1984: §50). It is possible "for the sickness of the philosophical problems to get cured," he tells us, but again, this can be accomplished "only through a changed thought and life" (Wittgenstein 1998, II, §23).

## 4.2 Cognitive Behavioral Therapy: "A Changed Thought and Life"

It can be challenging to modify our customs and traditions. We tend to fall into familiar patterns of thought and behavior. This is true of the philosophical life too.[3] Philosophers become set in their ways, employing customary tools of the trade to tackle issues. Therapeutic philosophy can be helpful in this endeavor by forcing philosophers to consider certain issues differently.

As mentioned previously, *Cognitive Behavioral Therapy* (CBT) has proven to be an effective form of treatment for a number of psychological issues. According to the American Psychological Association: "Numerous research studies suggest that CBT leads to significant improvement in functioning and quality of life."[4] Recall, in CBT the patient and therapist together work through a series of steps:

1. Identify troubling situations or conditions in your life.
2. Become aware of your thoughts, emotions and beliefs about these problems.
3. Identify negative or inaccurate thinking.
4. Reshape negative or inaccurate thinking.[5]

[3] A former colleague of mine, William Hirstein, once joked that as philosophers age they become afflicted with a "hardening of the categories."

[4] https://www.apa.org/ptsd-guideline/patients-and-families/cognitive-behavioral

[5] These steps, as well as what one can expect at each, can be found on the Mayo Clinic's website: https://www.mayoclinic.org/tests-procedures/cognitive-behavioral-therapy/about/pac-20384610

Adjusting the aims found in CBT to those of Wittgenstein's therapeutic method will not be exact; however, broadly speaking, both seek to effectively transition the "patient" from an unhealthy to a healthy condition. In therapeutic philosophy, this means discouraging investigations into *logically possible*, but at the same time, *humanly improbable*[6] statements, in favor of investigations based firmly in Common Sense. While reshaping unhealthy beliefs into healthy ones remains the primary goal, this can only begin if the philosopher *recognizes* that certain patterns of thinking he is familiar with are, in reality, flawed and wildly impractical. This recognition is crucial but can serve as the catalyst for change.

Let us briefly consider each of the four steps of CBT in relation to our therapeutic philosophy before turning to passages in *On Certainty* that support this therapeutic connection.

### 4.2.1 *Identify Troubling Situations or Conditions in Your Life*

In order to reshape unhealthy thinking, the patient must acknowledge and understand the circumstances that have produced his current mental state. In our case, the circumstances of this philosophical disease stem from acceptance of a flawed Cartesian methodology. When we encounter someone who *embraces* a practically un-tethered epistemology while simultaneously *rejecting* Common Sense, then philosophical therapy is recommended. When we encounter someone who permits—even *encourages*—the positing of logically possible but *impractical* questions, philosophical therapy is recommended.

The question is, how do we transition someone from this academically conventional way of thinking to something new? How are we to begin? Reshaping begins with this guiding question: what are the troubling philosophical questions and/or conditions we encounter in actually leading our lives?

### 4.2.2 *Become Aware of Thoughts, Emotions, & Beliefs About Problems*

Addressing the question above requires an awareness of one's philosophical thoughts and beliefs. Philosophizing within the Cartesian system may be where the ailing philosopher "comfortably functions," but he must be brought to see that what he *produces* within this system is wholly impractical. Oddly, a certain amount of *distrust* in what the patient produces can actually be healthy. Although distrust can

[6] Here, I simply mean it is incompatible with actual living or leading a human life. It is "humanly improbable" if it rules out indispensable human commitments or activities.

be intellectually frustrating, this frustration can also serve as a catalyst for closer inspection of one's beliefs. A closer inspection may lead to more tests. And, as stated before, if philosophy is to be a worthwhile and useful discipline, if it is to be more than merely thought experiments and mental exercises, it must work to serve *practical ends*.[7] This requires a "change in the sensibility of the patient." It is important he become aware when his philosophical questions begin exceeding the parameters of Common Sense inquiry.

### *4.2.3 Identify Negative or Inaccurate Thinking*

One of the more essential steps in this therapeutic process is getting the patient to comprehend what is wrong—"what is it I am suffering from?" Answering this question requires committed attention to why the thought, the question, the belief, etc. is "wrong," "negative," or "inaccurate." Philosophical therapy, much like CBT, is a slow process in which the philosopher is brought—by degrees—to a new understanding of the nature of the problem/s that trouble him.[8] Although the ultimate goal is to reshape the patient's thinking, we must not jettison his "diseased thought," instead, we must ensure it "run[s] its natural course" (Wittgenstein 1967: §382). Indeed, it took a lifetime of indoctrinated philosophical reasoning and inquiry to get the patient to this stage; hence, "slow cure" will be the best way to shake him free of what ails him. If done carefully, methodically, and unhurriedly, this "cure" will permit judicious reflection on the problem/s. Philosophical therapy requires patiently addressing the "diseased thought" while simultaneously applying new tools of understanding—tools that serve to demonstrate why the patient's past pattern of thinking was so damaging. And this leads us to the final step: *reshaping* the toxic thought.

It should be noted that this final step requires a willingness to "reshape" the old way of thinking. It may seem a more appropriate goal would be a willingness to *replace* the old way of thinking, but for Wittgenstein, it is imperative that the toxic thought never be completely removed, or better, never be completely forgotten. The old way of thinking ought never be completely blocked from the mind.

Reshaping one's old way of thinking means the patient will always keep the toxic thought close, but, if successful, the newly formed belief will prevent him from

---

[7] And in this way, it seems to me, Wittgenstein's (therapeutic) philosophy parallels the practicality of the American Pragmatists.

[8] On the psychological side of this analogy, although it does take time, CBT takes *less time* than say, Freudian psychotherapy.

slipping back into the unhealthy, old, impractical way of thinking.[9] Healthy, new, practical philosophical applications serve to *become* his "new normal."[10]

### *4.2.4 Reshape Negative or Inaccurate Thinking*

The tools leading to new "normal functioning" will include: a recognition on the philosopher's part that a distinction exists between ordinary skeptical questions (mistakes) and extraordinary ones ("mental disturbances"); and a recognition that to begin engaging in extraordinary doubt is the first step on a slippery-slope that leads to unhealthy, unfruitful (Cartesian) thought. Identifying the troubling situation, becoming aware of thoughts and beliefs, and recognizing negative or inaccurate thinking, are all necessary and important in curing the philosophical disease. But much like psychological therapy, in philosophical therapy, the patient must *want* to change. Philosophical maturation can only be achieved when the patient, alone, is capable of adjusting his/her behavior.[11]

## 4.3 Evidence of Therapeutic Philosophy in *On Certainty*

### *4.3.1 Identifying Troubling Situations or Conditions in Your Philosophical Life & Becoming Aware of Thoughts, Emotions, & Beliefs About Your Philosophical Problems*

Let us return to the guiding question posed above, modified from CBT: what are the troubling situations and/or conditions in your philosophical life? In *On Certainty*, the mistake/"mental disturbance" divide exposes the most obvious and troubling assumption that Cartesian skeptics make—the belief that inquiry can be extended to *anything*, short of logical contradiction, that all logical possibilities, regardless of probability, are equally worthy of consideration. Ironically, a troubling

[9] In the preface to *Philosophical Investigations*, Wittgenstein says he thought it might be helpful to publish his old thoughts from the *Tractatus* alongside his new ones here in the *Investigations*. As he says: "Four years ago I had occasion to re-read my first book (the *Tractatus Logico-Philosophicus*) and to explain its ideas to someone. It suddenly seemed to me that I should publish those old thoughts and the new ones together: that the latter could be seen in the right light only by contrast with and against the background of my old way of thinking." (Wittgenstein 1953: vi).

[10] This speaks directly to CBT's goal. Again, see the Mayo Clinic's website: https://www.mayoclinic.org/tests-procedures/cognitive-behavioral-therapy/about/pac-20384610

[11] In this way, the change is more accurately characterized by the more practical conception of therapy as cognitive (CBT), than by the often questionable conception of Freudian psychotherapy (involving the unconscious, childhood experiences, etc.). My thanks to Martin Benjamin for pointing this out.

inconsistency is exposed in Descartes' methodological framework right from the beginning, namely, the skeptic *assumes* there is someone there to receive his skeptical inquiries. Moore makes this very point in "Defense of Common Sense,"

> one way in which [skeptics] have betrayed this inconsistency, is by alluding to the existence of other philosophers. Another is by alluding to the existence of the human race, and in particular by using 'we' do so and so, e.g., that '*we* sometimes believe propositions that are not true,' is asserting not only that he himself has done the thing in question, but that *very many other human beings, who have had bodies and lived upon the earth*, have done the same (Moore 1925: 40).

Propositions about "other minds" are accepted as true, even if the skeptical philosopher has elsewhere pretended to deny such propositions as true. "The strange thing," Moore continues,

> is that philosophers should have been able to hold *sincerely*, as part of their philosophical creed, propositions inconsistent with what they themselves *knew* to be true; and yet, so far as I can make out, this has…frequently happened (Moore 1925: 41).

Wittgenstein also detects an inconsistency, his observation, however, cuts right to the heart of the matter: "[i]f you tried to doubt *everything* you would not get as far as doubting *anything*,… doubting itself presupposes certainty" (Wittgenstein 1972: §115). While the skeptic may believe his grounds for doubting are solid, he betrays himself by *beginning* with a belief—a belief that he *is*, in fact, doubting. But doubt of this magnitude does not even permit one to start, for it requires the skeptic to accept—to *believe*—that doubting is in fact what he is doing. The Cartesian skeptic never truly *begins* with doubt, for he *assumes* that he understands the word 'doubt' and what it refers to. At the bottom is a *belief* about doubt.

Again at §114, Wittgenstein identifies another glaring inconsistency in the Cartesian framework: "[i]f you are not certain of any fact, you cannot be certain of the meaning of your words either." His observation here upsets the Cartesian's most cherished, fundamental proposition—*Cogito Ergo Sum*. For to question *everything*, as Descartes claims he does, he would have to include questioning the very meaning of these words. But at no time does Descartes question whether this underlying proposition makes sense. Instead, he just *assumes* his internal language follows; he just *assumes* it has meaning.[12] From a therapeutic perspective, this appears quite

[12] This counterargument is the cornerstone to Wittgenstein's *private language argument* (PLA). The PLA occupies roughly §§243–315 in his *Philosophical Investigations*. It begins as follows: "But could we also imagine a language in which a person could write down or give vocal expression to his inner experiences-his feelings, moods, and the rest-for his private use?…The individual words of this language are to refer to what can only be known to the person speaking; to his immediate private sensations. So another person cannot understand the language." This passage serves to elicit a number of questions: *Could* this person's words refer to sensations? Is there a connection between the name and the *thing being named*? If so, that is, if there *is* a connection, *how* is it established? These questions all ask something similar, a question Wittgenstein himself poses in the very next passage: "how does a human being [say a child] learn the meaning of the names of sensations?-of the word "pain" for example?" Remember, this *would-be language* is completely private; its words are to pick out and to mean what *only the speaker can be aware of*-"objects" in the mind (or brain) such as pains, feelings, or images, that he, the child alone, is conscious of. Is

"troubling,"[13] and if our patient is being honest, it should cause some intellectual discomfort. As Wittgenstein once remarked in discussion with his friend, O.K. Bouwsma, all the Cartesian could really utter here is "Ah!" (Bouwsma 1986: 13–14).

Before we continue, let us take a moment to note what Wittgenstein is *not* saying about skepticism. He is *not* saying that doubt is *un*important, indeed, a certain amount of skepticism—in any discipline—is not only important, it is highly *encouraged.* A *healthy* dose of skepticism can prove helpful and sometimes even assist in propelling a discipline forward. (Consider if Copernicus, for example, had *not* questioned the Earth's place in the cosmos!) But the issue is when skepticism is applied *improperly*, that is, when one engages in *extraordinary* questions and answers. These are the sorts of skeptical questions so astonishing to anyone outside the discipline they might believe the person is mad. (Recall Wittgenstein's tree example above, "'This fellow is not insane. We are only doing philosophy.'") Radical skepticism of this sort prevents us from moving the discipline forward; we become paralyzed by it. This philosophical endeavor has proved to be *un*healthy.

After we have identified this unhealthy skepticism and become aware of our thoughts about it, our next step requires labeling this disease for what it is: "negative," "wrong," "impractical." Let us now turn to a few passages in *On Certainty* where Wittgenstein calls out particular aspects of this "philosophical disease."

### 4.3.2 *Calling It Out: Identifying Negative or Inaccurate Thinking*

At §55 Wittgenstein asks, "So is the *hypothesis* possible, that all the things around us do not exist? Would that not be like the hypothesis of our having miscalculated in all our calculations?" This example nicely illustrates the problem with the Cartesian framework. How do we assess the *absence* of objects? Perhaps the question is meant to suggest that the objects around us are *illusory*? But to ask whether any objects in the world really exist is bizarre; perhaps even *more* bizarre than asking whether we have miscalculated *all* of our calculations, however, it is precisely

this "*picking out the meaning*" legitimate? Can we, without *any contact with others, assign* meaning? Wittgenstein thinks this cannot be done; in fact, he thinks it is *impossible*. He does not believe that the child can *turn his attention inward*, as it were, to *learn* what the term "pain" means. For even if we assume the child *does know* the sensation of pain *from his own case*, what does the child *really* know here? It seems "he knows only what *he* calls '*pain*,' not what anybody *else* does." Remember, this is a *private language* after all; how can he know what pain is *only from his own acquaintance with it*? If *this* were possible, and everyone knew pain *only from his/her own case*, no one could teach *anyone else* the meaning of the word "pain" since the word "pain" would be defined solely by you, that is, *your inner experience of pain*. But since it is *your* pain and since I do not have *access* to it (nor you to mine), how could I *ever* know if we define it the *same way*?

[13] Passage, §114 coupled with what he says at §115, will also help us in *reshaping our negative or inaccurate thinking*—our final stage in therapeutic philosophy.

the sort of hypothesis the Cartesian would ask us to entertain. And if we *were* to entertain the hypothesis, we can immediately see there is something wrong, something inaccurate. Relying on the very sensations that produced the doubt in the first place *already* makes these sensations suspect, so our senses are not to be trusted. Similarly, in mathematics, Wittgenstein realizes if your calculations were incorrect, you would necessarily have to depend on the certainty of *other* calculations to verify the initial inaccuracy. But how certain are you of *these* calculations? The inquiry thus devolves into an endlessly vicious cycle. In §30 Wittgenstein says:

> When someone has made sure of something, he says: 'Yes, the calculation is right,' but he did not infer that from his condition of certainty. *One does not infer how things are from one's own certainty.*

But this elicits the question: how then *does* one infer "how things are"?

Responding to this question requires care, for one could easily get pulled back into the sorts of radical, skeptical questions that delivered him here in the first place. Based on his statement, it seems clear Wittgenstein does not think certainty in calculation can be acquired solely by one's self. (This kind of certainty, acquired from within, is what Descartes was seeking and is loaded with problems. I discuss these problems elsewhere.)[14] Rather, this kind of certainty is acquired through the help of our community, what Wittgenstein calls, our *form of life*. Here we are able to infer *how things are*. Our *form of life* is what provides meaning for us.[15] This idea is continued at §47: "*This* is how one calculates. Calculating is *this*. What we learn at school, for example." It is wrong for us to think of certainty in calculation as

---

[14] As explained in *Exploring Certainty*, "We all possess this strange sense in which we know what *must* come next when we are asked to count according to a mathematical rule. We say that a certain step follows necessarily—but why *necessarily*? The very notion of one's response being correct would seem to require that the answer is, as Wittgenstein puts it, "in some *unique* way predetermined, anticipated" (Wittgenstein 1953: §188). It is as if "the steps are *really* already taken, even before I take them in writing or orally in thought" (Wittgenstein 1953: §188). "But what," asks Wittgenstein, "does the peculiar inexorability of mathematics consist in?" (Wittgenstein 1998: I §4)...The certainty with which we hold this conviction is the result of our *training*. Just as we have been trained to recognize certain numerical series in our *form of life*, so too has a pupil in another *form of life*. As a participant in *his form of life*, he has his own inherited background, a history he possesses for recognizing numerical patterns. As such, it would be incorrect for us to say that he is committing an error here" (Brice 2014: 76–77).

[15] I also discuss mathematical certainty in *Exploring Certainty*. "With the help of two linguists, Peter Gordon—a behavioral scientist at Columbia University—spent more than two months observing members of the Pirahã tribe, a group of roughly 200 people who live in the Lowland Amazonia region of Brazil. The Pirahã's counting system consists only of the words "hói" (falling tone) for "one," "hoí" (rising tone) for "two," and "baagi" for "many." Daniel Everett, in his book, *Language: The Cultural Tool*, says that still further studies have shown that "the Pirahã's have no number words, nor any concept of counting in their language. This is the only community in the world, so far as we know, that has been determined to lack any numbers at all, not even the number 'one'" (Everett 2012: 260). (Brice 2014: 78–79).

something we can do alone—"from one's own certainty."[16] Rather, certainty is partly social—embedded in a linguistic community—and inhabits our everyday exchanges. Certainty, as I discuss in Chap. 6, can be acquired two different ways: it is not only instinctual, it is also borne out in the non-ratiocinated, *habitual* things we do—our *actions*, our *deeds* (Wittgenstein 1972: §402). The latter is a product of our environment.

We've discussed the distinction between mistake and "mental disturbance" in §§71–74, but just before this series begins, in §70, Wittgenstein gives an example of what he means by "mental disturbance."

> For months I have lived at address A, I have read the name of the street and the number of the house countless times, have received countless letters here and given countless people the address. If I am wrong about it, the mistake is hardly less than if I were (wrongly) to believe I was writing Chinese and not German.

The idea here is clear. If he were mistaken about where he lives, it would be so outlandish, so bizarre to a reasonable person it might be on par with not knowing what language one was writing in (perhaps he is writing in Chinese and *not* in German?). This kind of error is not merely a mistake; it is worse than being "wrong" or "inaccurate." Wittgenstein is calling this out for even *describing* it as a "mistake." Labeling it a "mistake" does not really capture how egregious an error it is. As he says at §72, "[n]ot every false belief of this sort is a mistake."

What then is the difference between these two types of error? "Or," as he says at §73, "what is the difference between my *treating* it as a mistake and my *treating* it as a mental disturbance?" It seems the key difference here has to do with our ability to follow a rule. When one commits a cognitive mistake (logical, mathematical, linguistic, etc), attempts are made to adjust to the standards within that particular discipline—within the particular *language-game* one is engaged in. Participation in the language-game means, over time, one becomes aware of what is permissible and impermissible, what the goal is, and so on. This may require some assistance and thoughtful reminding when a mistake occurs, but this is all quite normal and describes how we learn the rules within our given linguistic community. Now contrast this with a "mental disturbance." With a "mental disturbance," there is none of this. *This* sort of error is "aberrant, strange, incomprehensible" (Stroll: 1994, 47). At §75, Wittgenstein offers the following example:

> Would this be correct: If I merely believed wrongly that there is a table here in front of me, this might still be a mistake; but if I believe wrongly that I have seen this table, or one like it, every day for several months past, *and have regularly used it*, that isn't a mistake?[17]

[16] It is also wrong for us to conceive of calculating as a timeless, changeless absolute. And at the end of §47, he makes clear: "Forget this transcendent certainty, which is connected with your concept of spirit." His point here is that it is wrong to think of certainty in some removed, abstract sense, it is also wrong to affiliate it with a Cartesian sense of mind (or "spirit").

[17] He makes this same point at §119: "But can it also be said: Everything speaks for, and nothing against the table's still being there when no one sees it? For what does speak for it?" Isn't it our *use* of it? That is, my leaning against it, perhaps my putting a glass of water on it.

Perhaps our eyes are playing tricks on us, maybe it is dark and we *think* we see a table, when in fact it is actually a large box or some other item. We would not have an issue calling this a "mistake." But if I've seen it and used it for months and months only to learn I was wrong the whole time, describing *this* as a "mistake" does not represent how very wrong I've been. This error seems much greater than a mere "mistake."

Wittgenstein develops this "mistake/mental disturbance" further at §79, where he jokes about verifying his own gender:

> That I am a man and not a woman can be verified, but if I were to say I was a woman, and then tried to explain the error by saying I hadn't checked the statement, the explanation would not be accepted.

Wittgenstein assumes the explanation—"I had not checked the statement"—would not be accepted because conviction about one's own gender is something we possess without much thought. And while we can understand what Wittgenstein is attempting to say here, gender identity we now know is not that simple. To be fair, Wittgenstein was writing in the late 1940s when such topics were rarely, if ever, discussed, perhaps even recognized, among most philosophers. But today, we simply do not equate one's gender with one's sex. Sex is not always determined from external genitalia; nor are gender and sex binaristic. Furthermore, gender and sex are not both viewed objectively (that is, from a subjective standpoint, some people simply wish to identify themselves one way rather than another).[18] Therefore, distinguishing a mistake from a mental disturbance is better illustrated with his table example than with one rife with the complexities of gender identity.[19]

In Chap. 3, we discussed the fact that Moore's justification for his propositional knowledge is woefully insufficient. When pressed, he could not provide evidence (grounds); instead, he simply said there were things he knew.[20] But without "the right ground for his conviction," as Wittgenstein says, Moore simply "does not know." He makes the case for this in §91.

> If Moore says he knows the earth existed, etc., most of us will grant him that it has existed all that time, and also believe him when he says he is convinced of it. But has he also got the right *ground* for his conviction? For if not, then after all he doesn't *know* (Russell).

---

[18] This example is also curious given Wittgenstein's difficult relationship with his own sexuality. Some interpreters of Wittgenstein, like W.W. Bartley, believe that Wittgenstein's tormented personality was due, in part, to the guilt and shame he himself felt in being homosexual. Others, however, like Ray Monk, in his *Ludwig Wittgenstein: The Duty of Genius*, dispute this view. "Wittgenstein was uneasy, not about homosexuality, but about sexuality itself. Love," says Monk, "whether of a man or a woman, was something he treasured. He regarded it as a gift, almost as a divine gift...Sexual arousal, both homo- and heterosexual, troubled him enormously. He seemed to regard it as incompatible with the sort of person he wanted to be" (Monk 1990: 585).

[19] My thanks to Alison Reiheld for her helpful comments here.

[20] How am I to prove now that 'Here is one hand, and here's another'? I do not believe I can do it...I could not tell you what my evidence is; and I should require to do this at least, in order to give you a proof...[However,] I can know things which I cannot prove; and among things which I certainly did know, even if (as I think) I could not prove them, were the premises of my...proof (Moore 1939: 148).

What is of particular interest here is the parenthetical reference to Bertrand Russell. While we cannot be certain what Wittgenstein had in mind here, a best estimation points to Russell's "five-minute old earth" example, where he argued that there is nothing logically impossible or inconsistent in asserting that the earth is but *five minutes old.*

> There is no logical impossibility in the hypothesis that the world sprang into being five minutes ago, exactly as it then was, with a population that 'remembered' a wholly unreal past. There is no logically necessary connection between events at different times; therefore nothing that is happening now or will happen in the future can disprove the hypothesis that the world began five minutes ago (Russell 1992: 159–160).

Russell's hypothesis may be improbable, but from a logical point of view, it is *possible.* That is, as long as Russell does not violate the law of non-contradiction,[21] anything goes. Assuming Russell avoids logical impossibilities, he is limited only by his imagination. So, within this framework, the hypothesis that the earth was created five-minute ago, while extremely *unlikely*, is nonetheless, *logically intelligible.*

Putting logic aside for a moment, let us consider what any reasonable, practical, Common Sense person would say to Russell's hypothesis. It seems likely a reasonable person would say he had not made a *mistake* in his calculations, rather, a reasonable person would charitably say that Russell's conception of reality was wildly abnormal. In *Philosophical Investigations*, Wittgenstein put it this way:

> The proposition 'The Earth has existed for millions of years,' makes clearer sense than 'The Earth has existed in the last five minutes.' For I should ask anyone who asserted the latter: 'What observations does this proposition refer to; and what observations would count against it?'—whereas I know what ideas and observations the former proposition goes with (Wittgenstein 1953: Part II, p. 221).

The person who seriously asserts that the earth is but 5 min old is not committing a "mistake," that is, he is not wrong, inaccurate, or in error, in fact, it is wrong of *us* to even classify this as an "error." This is precisely the kind of person Wittgenstein would diagnose as suffering from a "disease of thought" or a "disease of the understanding."

Although Wittgenstein is, once again, calling out this kind of philosophical disease, it also bears repeating what he is *not* doing. He is *not* undervaluing the importance of a good questioning mind; he is *not* undervaluing the importance that skepticism plays in our acquisition of knowledge. Instead, his reference to Russell here is meant to serve as a warning: skepticism can be taken *too far.* Just as the person who suffers from obsessive–compulsive disorder needs to know where the

[21] Russell does not ask us to imagine something that is logically *im*possible, e.g., something that is simultaneously black and white; a triangle that lacks three sides; a round square. We immediately rule out blatant contradictions like these because they are *un*intelligible to us. To Russell's credit, he follows this up with this acknowledgment: "I am not suggesting that the non-existence of the past should be entertained as a serious hypothesis. Like all skeptical hypotheses, it is logically tenable, but uninteresting."

line is, the philosopher too must know the boundary between reasonable and unreasonable doubt (between a mistake and a "mental disturbance").[22]

We can now turn to the final stage in our philosophical therapy, reshaping negative or inaccurate thoughts.

### *4.3.3 Mending Philosophical Thought: Reshaping Negative or Inaccurate Thinking Through Reasonable Inquiry & Common Sense*

As a point of entry, re-establishing a scope and limit to philosophical questions and answers seems like a reasonable place to begin. In a number of passages throughout *On Certainty* Wittgenstein provides a rough, first draft account of how we might go about setting this scope and limit of acceptable philosophical inquiry. Let us begin with the following cluster of passages: §§114, 115, 122, 160, and 143.[23]

As we noted above, at §114, Wittgenstein identifies a troubling inconsistency in the Cartesian framework: "[i]f you are not certain of any fact, you cannot be certain of the meaning of your words either." A moment's reflection reveals just how profound this statement is. Descartes believed he could avoid the most extreme aspects of radical skepticism by accepting the meaning of his words: *Cogito Ergo Sum*. But, if *everything* is subject to doubt, "if you are not certain of any fact," you should also have to doubt the very meaning of your words or, minimally, doubt whether the meaning of your words retain their same meaning from one utterance to the next. With this remark, Wittgenstein has done more than call out the "philosophical disease," he has found its root cause. He follows up this devastating observation at §115, with this:

[22] In the opening passage of §66, Wittgenstein mentions the "different degrees of assurance" he has about reality. "How does the degree of assurance come out," he asks. "What consequences has it?" Let us start with what Wittgenstein means by "assurance." "Assurance," he says, he means "being sure of something," but even though "I may be sure of something," "I may…still know what test might convince me of error. I am e.g. quite sure of the date of a battle, but if I should find a different date in a recognized work of history, I should alter my opinion, and this would not mean I lost all faith in judging." This "recognized work of history" provides him some guarantee (some "degree of assurance") against what he *thought* he knew. This is the "test" that "convinces [him] of error." But this error does not mean he begins questioning *everything*; he would not lose "all faith in judging" if this were to occur. Indeed, we all make these sorts of mistakes, but we do not begin wondering if anything *at all* can be trusted. So why does Wittgenstein jump to the most extreme case: "all faith in judging" *could* be lost? The reason has to do with the Cartesian Method of Doubt. Descartes made it permissible to push "judging" to absurd levels, but here, Wittgenstein is pushing back.

[23] As this scope is a reaction to the very broad Cartesian scope, at §383 Wittgenstein offers yet another argument illustrating the weakness of Cartesian skepticism: "The argument 'I may be dreaming' is senseless for this reason: if I am dreaming, this remark is being dreamed as well—and indeed it is also being dreamed that these words have any meaning."

> If you tried to doubt everything, you would not get as far as doubting anything. The game of doubting itself presupposes certainty.

There are two significant and related points to unpack in this passage: first, this kind of doubting—philosophical doubting—takes place in a *language-game*; second, Wittgenstein has cleaved a categorical distinction here between language-games on the one hand, and *certainty* on the other. Let us briefly address each of these in turn.

As discussed in Chap. 1 (and more fully in the Appendix), language-games are "part of an activity, or…*form of life*" (Wittgenstein 1953: §23). They emphasize the wide variety of our uses of language, but, like games, they are also rule-guided activities. "The game of doubting," as Wittgenstein calls it, comprises just one of our many different language-games. Doubt is a tool that can be applied in a variety of different language-games, however, since Wittgenstein's focus throughout *On Certainty* has been on epistemic issues, it seems he wants us to look at doubt as a tool used in the language-game of knowledge.

The other significant point at §115 is Wittgenstein's categorical division between language-games and *certainty*. To doubt something—*anything*—means you must *accept* something first. Here, at §115, he begins to realize that certainty must be in place first and foremost *before* we begin analyzing what occurs in any of our language-games. It is certainty that lies at the bottom of this (and every other) language-game. But what, exactly, does he *mean* by "certainty"? How is *certainty* analyzed? We will explore these topics in more detail in Chap. 6. For now, let us consider a few other passages that illustrate how these two points at §115 are related *and*, more importantly, how they connect to our therapeutic process.

At §122, Wittgenstein poses the following question: "Does not one need grounds for doubt?"[24] And a natural follow-up question to this seems to be: "If so, what *are* those grounds?" At §160, he offers a partial answer: "The child learns by *believing* the adult. Doubt comes *after* belief." What Wittgenstein has not yet clearly stated

[24] The American pragmatist, C.S. Peirce, makes this same point, in both "Some Consequences of Four Incapacities" and "The Fixation of Belief." Peirce argues, if and when we doubt, we must have a "positive reason" to do so. An inquirer possesses a body of settled beliefs—beliefs which are, in fact, *not doubted*—however, these beliefs are susceptible to doubt if prompted by some "positive reason." Peirce, in contradistinction to Descartes, says, just because we *can* pose a skeptical question does not make it a worthwhile question. What matters is whether it is "real and living," that is, whether it is meaningful to our lives. "[T]he mere putting of a question into the interrogative form does not stimulate the mind to any struggle after belief. There must be a real and living doubt and without this all discussion is idle" (Peirce 2011a, b: 11). Cartesian global doubt is not genuine doubt at all; it is an attempt to see how far one can logically push skeptical inquiry. We either have a reason for doubting, says Peirce, or we do not. If we have a reason to doubt, this is a genuine reason. If we doubt in a way that Descartes doubts, we have, what Peirce calls, 'fictitious doubt,' for it serves no purpose other than pushing it to extreme heights. Cartesian doubt is not genuine doubt. When skepticism is being used but it serves no practical end, this is an erroneous conception of a proof that ought to be swept away. As Peirce says: "Some people seem to love to argue a point after all the world is fully convinced of it. But no further advance can be made. When doubt ceases, mental action on the subject comes to an end; and, if it did go on, it would be without a purpose" (Peirce 2011a, b: 11). Doubt, therefore, must *serve a purpose*. It does not serve to further the discourse if we are doubting simply because we *can*.

(perhaps it is not yet clear in his mind) is that the ground here—the "belief" here—is *non-ratiocinated*. That is, the grounds represent a conviction that is *not at the level of conscious thought*. Convictions here might include such beliefs as, books have pages, heavy objects tend to fall to the ground when no longer held, fire is hot, triangles have three sides, there are other minds, etc. These sorts of convictions are so basic, so fundamental that, after we learn them at an early age, they rarely rise to the level of conscious or articulated thought (except, perhaps, in philosophy seminars!).[25] It is helpful to remember the aim here in the final step of the therapeutic process: by re-establishing a sensible scope and limit our philosophical questions and answers, we seek to reshape negative or inaccurate philosophical thoughts. This final step has an additional benefit of bringing Common Sense convictions along with it. Let us consider some evidence for this.

In passages §§219–221, Wittgenstein returns to his division of doubt between a mistake and a "mental disturbance," however, this time he explicitly uses the phrase, "reasonable person," which can only properly be contrasted with "unreasonable person."

> §219 There cannot be any doubt about it for me as a reasonable person—That's it.—
> §220 The reasonable man does *not have* certain doubts.
> §221 Can I be in doubt *at will*?

While there are some doubts that this reasonable person simply *does not have*—some things this reasonable person *cannot be mistaken about*—it would obviously be wrong to say the reasonable person is *never* mistaken. Of course, he doubts *some* things, and he may be right to doubt them, but he may also be *mistaken* about some of these things. One conviction the reasonable person cannot doubt, Wittgenstein says, includes his never having been in the stratosphere: "I cannot possibly doubt that I was never in the stratosphere." Wittgenstein follows up this conviction with questions like: "Does that make me know it? Does it make it true?" With our therapeutic process in mind, let us consider how we might answer these sorts of questions.

There are countless propositions—basic propositions—that we rarely think about; there are even more, however, that we *never* think about. When those rarely thought of, basic propositions *are* brought to the level of contemplation and we *do* consider them, it seems rather trivial to utter them. And while they may be trivially true, it seems a mischaracterization to say that we "know" them. Rather, they are, as Wittgenstein puts it, "there, like our life" (Wittgenstein 1972: §559). *This* is the sort of Common Sense that simply comes along in our reshaping effort.

Finally, let us return to one passage from our group of passages, namely, §221. "Can I be in doubt *at will*?" How might we respond to Wittgenstein? Can you be in doubt *at will*? He does not mean *if the context is right*, e.g., did I turn off the stove

[25] Paradoxically, much to Descartes' chagrin, it appears *belief* provides us with grounds for doubt. As Wittgenstein says at §143: "A child learns there are reliable and unreliable informants much later than it learns facts which are told it. It does not learn *at all* that that mountain has existed for a long time: that is, the question whether it is so does not arise at all. It swallows this consequence down, so to speak, together with *what* it learns."

when I left the house? Did I feed the meter? Rather, he means can one, at *any time*, be in doubt *at will*? Wittgenstein clearly thinks the answer here is so obvious it does not deserve a response. But *had* he replied to his own question, he would have probably said something like, but is this *really* doubt? Are we *really* experiencing a mental state of doubt when we question whether we exist, for example? Is *this* the kind of doubt that a reasonable, Common Sense person engages in? Is *this* the kind of doubt on which anything that really matters to us, hangs? Clearly not. This kind of doubt is aberrant and sounds more like a *mental disturbance*. It certainly is not a "mistake." Yet, this is precisely what philosophy has, for centuries, willfully encouraged.

In order to help our patient through recovery, it is important that he be able to identify the point at which inquiry passes from reasonable to unreasonable doubt. Any reshaping that takes place here assumes we have already been working with him, patiently assisting him through the previous three therapeutic stages.

## 4.4 Therapeutic Philosophy: A Reshaping of the Tradition

As we have seen, Wittgenstein's *therapeutic philosophy* is a prescription for a slow but steady readjustment in how the patient confronts traditional philosophical problems. This new course of action comes in stages:

- *Acknowledging and understanding the circumstances that have produced this current mental state.*

In our case, the circumstances of this philosophical disease stem from acceptance of a flawed Cartesian methodology.

- *An awareness of one's philosophical thoughts and beliefs about the problem/s.*

As we have seen, philosophizing within the Cartesian system does not serve *practical ends,* nor does it serve anything that really matters to us—that is, what matters in real life, not just what sometimes devolves into a parlor game in philosophy classes or discussions. This means becoming aware of when philosophical questions begin to exceed the parameters of Common Sense.

- *An essential step in this therapeutic process—getting the patient to identify and comprehend why his thinking was wrong, inaccurate, or negative.*

Philosophical therapy requires addressing the "diseased thought" while simultaneously applying tools that serve to demonstrate why the patient's past pattern of thinking was so damaging. In our case, this means accepting Common Sense truisms and setting a scope and limit on acceptable forms of inquiry.

- *The final step—reshaping the negative or inaccurate thought/s—requires applying tools in an effort to institute new "normal functioning."*

These tools include: a recognition on the philosopher's part that a distinction exists between ordinary skeptical questions (mistakes) and extraordinary ones (mental disturbances); and a recognition that extraordinary skeptical doubt often leads one to unhealthy, impractical thought.

## References

Bouwsma, O.K. 1986. *Wittgenstein Conversations 1949–1951*, ed. J.L. Craft and Ronald E. Hustwit. Indianapolis: Hackett.
Brice, Robert Greenleaf. 2014. *Exploring Certainty: Wittgenstein and Wide Fields of Thought*. Lanham: Lexington Books.
Everett, Daniel. 2012. *Language: The Cultural Tool*. New York: Pantheon Books.
Monk, Ray. 1990. *Ludwig Wittgenstein: The Duty of Genius*. New York: The Free Press.
Moore, G.E. 1909. Hume's Philosophy. In *Philosophical Studies*. London: George Allen and Unwin Ltd.
———. 1919. External and Internal Relations. In *Philosophical Studies*. London: George Allen and Unwin Ltd.
———. 1925. Defense of Common Sense. In *Philosophical Papers 1962*. New York: Collier Books.
———. 1939. Proof of An External World. In *Philosophical Papers 1962*. New York: Collier Books.
———. 1952. Certainty. In *Philosophical Papers 1962*. New York: Collier Books.
———. 1953. What is Philosophy? In *Some Main Problems of Philosophy*. New York: Collier Books.
———. 1993. (Originally published 1903). *Principia Ethica*. Cambridge: Cambridge University Press.
Peirce, C.S. 2011a. The Fixation of Belief. In *The Philosophical Writings of Peirce*. New York: Dover Publications.
———. 2011b. Some Consequences of Four Incapacities. In *The Philosophical Writings of Peirce*. New York: Dover Publications.
Russell, Bertrand. 1992. (Originally published 1921). *The Analysis of Mind*. London: Routledge.
Stroll, Avrum. 1994. *Moore and Wittgenstein on Certainty*. Oxford: Oxford University Press.
Wittgenstein, Ludwig. 1953. *Philosophical Investigations*. Trans. G.E.M. Anscombe. Oxford: Blackwell Publishers.
———. 1967. *Zettel*, ed. G.E.M. Anscombe and G.H. von Wright, and trans. G.E.M. Anscombe. 2nd ed. Oxford: Blackwell Publishers Ltd.
———. 1972. *On Certainty*, ed. G.E.M. Anscombe and G. H. Von Wright, and trans. D. Paul and G.E.M. Anscombe. New York: Harper & Row.
———. 1978. *Philosophical Grammar*, ed. R. Rhees, and trans. Anthony Kenny. Los Angeles: University of California Press.
———. 1984. *Culture and Value*, ed. G.E.M. Anscombe and G.H. von Wright, trans. Peter Winch. Chicago: University of Chicago Press.
———. 1992. *Last Writings On The Philosophy of Psychology*, Vol. II, ed. G.H. von Wright and Heikki Nyman, trans. C.G. Luckhardt and Maximilian A.E. Aue. Oxford: Blackwell Publishers.
———. 1998. *Remarks on the Foundations of Mathematics*, ed. G.E.M. Anscombe, G. H. Von Wright and Rush Rhees. Cambridge: MIT Press.

# Chapter 5
# Knowledge and Belief

**Abstract** This chapter provides an analysis of Wittgenstein's rejection of the conventional understanding of the relationship between knowledge and certainty, and a reshaping of our traditional conception of belief.

**Keywords** Action · Belief · Certainty · Cognitive Behavioral Therapy (CBT) · Framework reading · Hinge-proposition · Justified · Knowledge · G.E. Moore · True · Ludwig Wittgenstein

## 5.1 "In the Beginning Was the Deed"

After more than four hundred remarks in *On Certainty*, Wittgenstein quotes a line from Goethe's *Faust*: "In the beginning was the deed" (Goethe 1962: Line 1237).[1] In Goethe's novel, the quote stands in opposition to the *New Testament* verse: "In the beginning was the word" (John 1:1). Despite the religious context, Wittgenstein's decision to quote it here is appropriate. By moving away from thinking of certainty strictly in propositional terms towards thinking of it as something actional, his efforts in *On Certainty* align nicely with Goethe's quote. Indeed, as Ray Monk observes, "[t]his quote could be regarded as the motto of *On Certainty*" (Monk 1990: 579).

From a therapeutic standpoint, however, the quote works for a very different reason. Goethe had his character, Heinrich Faust, preface the line with a bit of self-assurance, telling himself to "write with confidence"[2] as he begins rewriting verses from the *Gospels*. Given the radical reconstruction of ideas Wittgenstein is developing in *On Certainty*, he *too* needs confidence as he struggles against long established and widely held beliefs in philosophy. He *too* must have faith in his own abilities as he begins reshaping and restructuring ideas that have become "gospel

[1] "Im Anfang war die Tat." Wittgenstein cites Goethe in *On Certainty* at §402.

[2] "...und schreib getrost..."



R. G. Brice, *Wittgenstein's On Certainty: Insight and Method*, SpringerBriefs in Philosophy, https://doi.org/10.1007/978-3-030-90781-5_5

truth," so to speak, to traditional, academic philosophers. In this chapter, we will consider some of these beliefs, beginning with the classic definition of knowledge as *justified true belief.*

## 5.2 Justified True Belief as Knowledge

Philosophers have been evaluating the requirements necessary for a person to claim that they "know something" at least since Plato first raised the issue in his dialogue the *Theaetetus*. There, knowledge was defined as "true belief combined with a *logos*," or "justification" (Plato 1961: St. 201c–210b). Knowledge was thus characterized *justified true belief* (JTB).[3] All three conditions—justification, truth, and belief—were required in order for us to claim that we knew something. Let's briefly consider these conditions, beginning with *belief.*

For Plato and much of the Western philosophical tradition, belief could be about almost anything. But in order to convert belief into something normatively *believable*—something that *ought* to be believed—one needed to demonstrate that it was not only true, but that there was some compelling evidence to support it. If, for example, a person believes that hurricanes are increasing in frequency and intensity because of climate change, and later reads a few scientific journals that provide evidence in support of her conviction, her belief has now been *strengthened*, she is now prepared—*with evidence*—to back it up. And while *justification* of a belief is necessary, it remains insufficient to transform it as knowledge. For this to occur, her justified belief must also be *true*. Let's consider some examples.

The history of science provides many examples where a belief is justified at a certain point in time, but this *justified belief* later proves to be false. For example, many people once thought they were justified in believing that:

- the Earth is flat;
- we are located at the center of the universe;
- human beings are categorically distinct from other animals;
- a fire-like element (phlogiston) is present in combustible bodies.

The missing (and necessary) condition in each of these is *truth*. Absent this key component, we cannot say *justified belief* constitutes knowledge.

---

[3] Sometimes called the "tripartite analysis of knowledge," it is also referred to as the "traditional analysis." This definition has been around for a very long time: the JTB principle was first introduced by Plato (despite the fact that Plato himself rejected it in favor of his theory of Forms) most fully in his dialogue the *Theaetetus*. Theaetetus' definition of knowledge is that knowledge—*episteme*—is "true opinion [*alethos doxa*] combined with reason [*logos*]." St. 201c–d. We find the first germs of this idea in another of Plato's dialogues, the *Meno*, St. 97a–100b, where he says that true belief can be converted into knowledge when one gives an account (*logos*) of it. This is also found in one other Platonic dialogue, the *Symposium*, St. 202a.

Nor would it be correct, for that matter, to say that *true belief* is sufficient for knowledge. In order to understand why this is the case, imagine a clock that is no longer keeping time, having stopped with both hands pointing directly at the 12. Unaware that the clock is broken, and by chance, let's say you look at it *precisely* at twelve o'clock. As a result, you believe it *is* twelve o'clock. So you not only *believe* it is twelve o'clock, your belief is actually *true*. But we do not call this *knowledge* because of the *way* in which this information was attained—it was simply coincidence that the correct time was twelve o'clock. *Coincidental* true belief is not knowledge. What this true belief lacks is reliable *justification*. As such, all three conditions—*justification, truth*, and *belief*—must be present to constitute a genuine knowledge claim.

## 5.3 Conflating Belief with Knowledge

"Here is one hand, and here is another" (Moore 1939: 144). Recall in "Proof of an External World," Moore offered this basic proposition as support for his claim that there are objects external to us. Later in the article, he claimed to "know…the premises of [his]…proof," to know that here is one hand, etc., despite the fact that, as he readily acknowledged, he "could not prove them" (Moore 1939: 148).

*On Certainty* begins with an obvious reference to Moore: "If you do know that *here is one hand*, we'll grant you all the rest" (Wittgenstein 1972: §1). Tacitly, and in his own way, Wittgenstein poses a version of *therapeutic philosophy's* guiding question here: what are the troubling situations and/or conditions in your philosophical life? Wittgenstein's implicit question might be put this way: Can you demonstrate for me that Moore is correct? For if Moore's claim to knowledge is beyond question, if his claim is indeed, beyond doubt, then *everything* else that stems from it will be accepted—"we'll grant you all the rest." However, if there *is* a reason to doubt it then it would be incorrect to say he *knows* it. For this would mean that one (or more) of the characteristics that produced the knowledge—justification, truth, or belief—has failed. We would thus be using the term incorrectly. This is what Wittgenstein thinks occurs with Moore. For as he later laments, "we just do not see how very specialized the use of 'I know' is" (Wittgenstein 1972: §12).

Use of the phrase, "I know…"—a shortened version of "I know that *p*"—is what philosophers refer to as *propositional knowledge* (or "knowing that").[4] Propositional knowledge is the kind of knowledge that occurs when a person says she knows *that* a proposition is true. For example, "Whales are mammals" is a proposition, while "Cecelia *knows that* whales are mammals" is an attribution of propositional knowledge. As we saw above (5.2), when philosophers use this concept, it requires that the belief be justified and true. This, however, is *not* the way Moore uses it. What Moore

[4] "Knowing *that*" is sometimes contrasted with "knowing *how*" (Ryle 1949: 28–29). Passages §§564–566 in *On Certainty* point to Wittgenstein's division between "knowing that" and "knowing how."

seems to be offering with "I know that I have two hands" is more like a guarantee—a guarantee that what he *knows* is an unquestionable, indubitable fact. But this assurance only reveals the strength of his *conviction*, it only reveals his *belief* that he knows, not that he does indeed *know*. So when Moore says, "I know that I have two hands," but then fails to offer justification, this is really no different than him offering his *belief* about it.

At §520 and §521, Wittgenstein discusses Moore's misuse of "I know," specifically, his failure to include (any) justification:

> §520 Moore has every right to say he knows there's a tree there in front of him. Naturally he may be wrong. (For it is *not* the same as with the utterance "I believe there is a tree there.") But whether he is right or wrong in this case is of no philosophical importance. If Moore is attacking those who say that one cannot really know such a thing, he can't do it by assuring them that he knows this and that. For one need not believe him. If his opponents had asserted that one could not *believe* this and that, then he could have replied: '*I* believe it.'
>
> §521 Moore's mistake lies in this—countering the assertion that one cannot know that, by saying 'I *do* know it.'

Use of the phrase "I know…" doesn't *end* a demand for justification, it *invites* it. "What is the proof that I *know* something?" a frustrated Wittgenstein asks. "Most certainly not my saying I know it" (Wittgenstein 1972: §487). Yet Moore's response to, "How do you know?" is to insist that there *are* things he knows (in some cases, even things he knows *with certainty*)[5] before subsequently admitting that he is unable to provide any reason for these things. But as Wittgenstein explains,

> If someone believes something, we needn't always be able to answer the question 'why he believes it'; but if he knows something, then the question "how does he know?" must be capable of being answered (Wittgenstein 1972: §550).

In short, Moore's error consists in conflating very secure Common Sense belief or conviction with propositional knowledge. As a result, he fails to see the legitimacy of a demand for justification.

Here, and in Chap. 3, we discussed the circumstances that produced Moore's mistake. We analyzed and assessed why and how his argument goes wrong. The rules that exist within our *language-games* and within our *form of life* proved valuable tools here. Specifically, our application of rules within the *language-games* of logic, epistemology, and metaphysics helped us evaluate where Moore's argument went wrong and, perhaps more importantly, *why*. We applied these tools while simultaneously refraining from exceeding any Common Sense parameters. Despite this, philosophy—or the *language-games* that make up the discipline—remain limited in that philosophy has nothing to say about *awareness* of our thoughts, emotions, and beliefs. *Philosophical therapy*, however, can play a role here. For in

[5] In reference to his common sense propositions, Moore says, "I have nothing better to say than it seems to me that I do know them with certainty" (Moore 1925: 41).

addition to understanding the circumstances and evaluating the soundness of arguments, philosophical therapy counsels us to be *aware* of our thoughts, emotions, and beliefs about such matters.

In Chap. 3, we discussed CBT as a model for philosophical therapy. Recall, with CBT treatment, one of the first things the therapist must do is unearth what lies beneath the patient's beliefs and behaviors. Once the problem has been identified, the therapist then attempts to alter what the patient takes to be "normal functioning" and reshape it into something *new*. When successful, this new functioning ultimately *becomes* normal; the hope is to establish a deep-seated change in the person. In this way, therapy can effect a change in the *sensibility* of the patient. Effecting this change occurs over time, during which patient and therapist engage in a discursive back and forth. In time, the patient gradually comes to recognize what his hidden patterns of beliefs are, why they are destructive, and how (with the help of the therapist) he can develop tools to manage and function whenever these beliefs and behaviors re-emerge.

In numerous passages in *On Certainty*, Wittgenstein transcribes an internal dialogue in which he plays the role of both therapist *and* patient. This is rather complicated role-playing, for while trying to unearth hidden patterns of beliefs that the Cartesian holds, Wittgenstein must also uncover patterns of belief in *himself* that have thus far remained concealed. Determining *what* these beliefs are is only the first step; he also must express to the patient (himself) *why* they are destructive and *how*—through a persistent, internal dialogue—he can help develop tools to begin managing and functioning properly. Here are just a few of the many passages[6] where this complicated role-playing is on display:

§133 Under ordinary circumstances I do not satisfy myself that I have two hands by seeing how it looks. *Why* not? Has experience shown it to be unnecessary?…

§150 …How do I know that this color is blue? If I don't trust *myself* here, why should I trust anyone else's judgment? Is there a why? Must I not begin to trust somewhere? That is to say: somewhere I must begin with non-doubting; and that is not, so to speak, hasty but excusable: it is part of judging.

§347 "I know that that's a tree." Why does it strike me as if I did not understand the sentence? though it is after all an extremely simple sentence of the most ordinary kind? It is as if I could not focus my mind on any meaning…

§407 For when Moore says, "I know that that's a…" I want to reply "you don't *know* anything!—and yet I would not say that to anyone who was speaking without philosophical intention. That is, I feel (rightly?) that these two mean to say something different.

§409 If I say, "I know that that's a foot"—what am I really saying? Isn't the whole point that I am certain of the consequences…Would my knowledge still be worth anything if it let me down as a clue in action? And *can't* it let me down?

§498 The queer thing is that even though I find it quite correct for someone to say "Rubbish!" and so brush aside the attempt to confuse him with doubts at bedrock,—nevertheless, I hold it to be incorrect if he seeks to defend himself (using, e.g., the words "I know").

§515 If my name is *not* L.W., how can I rely on what is meant by "true" and "false"?

[6] Other passages include: §§6, 32, 137, 150, 307, 435, 481, 495, 553.

Wittgenstein not only exposes hidden patterns of belief that the Cartesian skeptic regards, he also discovers hidden beliefs he *himself* holds. Developing proper tools to overcome the "disease" will take time. Treatment occurs gradually here, but a "slow cure" is likely to insure a successful outcome.[7]

## 5.4 From "Hinges" to *Action*: Reshaping Our Conception of Belief

The notion of belief that Wittgenstein eventually discovers in *On Certainty* centers on a class of convictions that are so fundamental we rarely stop to consider whether they are true or false; we rarely stop to consider whether justificatory reasons can (or even should) be provided.[8] These convictions, he says,

> simply get assumed as truism[s], never called into question, perhaps not even ever formulated. It may be for example that *all inquiry on our part* is set so as to exempt certain propositions from doubt, if they are ever formulated. They lie apart from the route traveled by inquiry (Wittgenstein 1972: §§87–88).

If, however, someone *were* to supply reasons for a fundamental conviction, the banality of the statements might cause us to question this person's sanity. At this stage in his exploration, Wittgenstein, like Moore, thinks some of what he considers propositions are "such obvious truisms" they are not "worth stating" (Moore 1925: 32). Unlike Moore, however, he recognizes the skeptical problem that emerges when formulating these basic beliefs as propositional knowledge claims.

> [H]ow do I *know* that it is my hand?…When I say "how do I know?" I do not mean that I have the least *doubt* of it. What we have here is a foundation for all my action. But it seems to me that it is wrongly expressed by the words "I know" (Wittgenstein 1972: §414).

If we contrast this view with the way much of the Western tradition has conceived of belief, we find that the kinds of convictions Wittgenstein is becoming explicitly aware of here constitute a very special class. Belief of this sort is not an ingredient in knowledge, it "lie[s] apart" from it. He likens it to hinges fixed on a frame (Wittgenstein 1972: §§341–343). "If I want the door to turn," says Wittgenstein, "the hinges must stay put" (Wittgenstein 1972: §343). These "hinge propositions," as they have come to be known, stay put so that our thoughts and practices are possible. However, it is important to note that the "propositional" nature of the "hinge" is really just a confusing vestige of an earlier and misleading way of thinking—a way Wittgenstein is slowly coming to grips with. Gradually, the hinge's identity as first and foremost *propositional* drops out. These certainties are more accurately characterized as unreflective beliefs or convictions *embedded in*

---

[7] Note here the similarity with ancient Greek philosophy, especially Plato. Platonic dialogues take time (and patience). Socrates displays ample patience in dealing with ("treating") mistaken discussants like Euthyphro, Crito, and others. It's a forerunner of Wittgenstein's "slow cure."

[8] This goes equally for someone like the skeptic—who doesn't affirm but *doubts* these basic beliefs.

*action*; they assume a sort of counterfeit "propositional form" only when we *need* to talk about them, only when we need to identify them, *linguistically*. From a therapeutic standpoint, it is the "slow cure" that helps distinguish actional certainties from genuine propositions. Although they usually go unnoticed, the conduct of our lives demonstrate that we unreflectively "accept" them (Wittgenstein 1972: §344). Curiously, however, Wittgenstein also says that they are "exempt from doubt" (Wittgenstein 1972: §341).

Given the "very specialized use of 'I know'" and the problems Moore's misuse of it revealed, Wittgenstein understandably avoids couching hinge-propositions inside knowledge claims. Yet, his declaration that they are "exempt from doubt" seems dubious. At this stage, hinge-propositions are *propositional*, and the skeptic, upon hearing someone utter a basic statement like, "The earth exists," will not be dissuaded from inquiring, "Yes, but *how* do you know?" To spare hinge-propositions from doubt in this way, appears as careless as Moore[9] insisting that his common sense propositions are known, "even if [he] could not prove them" (Moore 1939: 148). So is Wittgenstein being careless here? Why would he say such a thing?

Some scholars, like Annalisa Coliva, contend that Wittgenstein's move here is perfectly acceptable. Hinges are exempt from doubt, she says, because "hinges are rules," and rules

> ...can't be assessed in terms of truth and justifiedness or even knowledge...[S]kepticism raises a doubt where doubt cannot rationally be sustained. Therefore, skeptical doubts are nonsensical but not because they are meaningless but because they are raised where it makes no sense to raise them and are therefore *not rational* (Coliva 2016: 11).

On what she and Daniele Moyal-Sharrock have called the "framework reading" of *On Certainty*, hinge-propositions "behave like rules" because "they are required in order to collect evidence for or against ordinary empirical propositions. What that means," she says, "is that they are rationally mandated" (Coliva 2016: 17). For support here, Coliva turns to a key passage:

> §342 That is to say, it belongs to the *logic* (her emphasis) of our scientific investigations that certain things are *in deed* [*in der Tat*] not doubted.

"Here," she says,

> Wittgenstein...is making clear that it belongs to the logic—i.e., to the norms that regulate our investigations—that certain propositions aren't doubted...Hence, hinges must stay put, and thus behave like rules (Coliva 2016: 17).

Although not entirely compatible with our therapeutic interpretation of *On Certainty*, the *framework reading* has some merit. Wittgenstein's division between a mistake and a "mental disturbance" presupposes some sort of rule must be in place in order to have us label doubts about hinge-propositions as "mentally disturbed." Indeed, questioning a hinge-proposition is a key indicator that the individual posing the questions is, perhaps, unhealthy; as Martin Benjamin aptly put it: people who pose such questions have come "unhinged" (Benjamin 2003: 63). To that extent, the

[9] Perhaps even *more* careless, given that Wittgenstein had read Moore!

*framework reading* seems correct; for if there were no rule in place, it would be difficult to offer an assessment.

Yet, we cannot ignore what Wittgenstein says about hinges as rules. At §117, while contemplating why it is impossible for him to doubt that he has never been on the moon, he says:

> [f]irst and foremost, the supposition that perhaps I have never been there would strike me as *idle*. Nothing would follow from it, nothing be explained by it. It would not tie in with anything in my life. When I say "Nothing speaks for, everything against it," this presupposes a principle of speaking for and against. That is, I must be able to say what *would* speak for it.

The very idea that the meaning of our hinge-propositions is fixed *prior* to their being used is what Wittgenstein explicitly criticizes at §117. And elsewhere, he has stated his rejection of this quite bluntly, "[t]he rule, not the exception, is the incurable illness" (Wittgenstein 1982: §110).

What Coliva's emphasis on the "logic" (or rules) aspect in §342 overlooks is the much more important point occurring in the passage, indeed, the very point Wittgenstein *himself* emphasizes, "certain things are *in deed* not doubted."[10] By "deed," he means, "practice," "action," the "things we *do*." That is to say, "…certain things are *in practice* not doubted." Coliva's case for rules could be made stronger if, as I argue in Chap. 6, she called attention to the fact that some propositions *become* certain, passing from a propositional, ratiocinated, rule-based *language-game*, down to—*and becoming*—an unreflective, non-propositional actional certainty. The proposition "arrives at" this actional certitude through positive reinforcement of the rules. These are rules that become hinges *to the individual*; but these are not hinges *as* rules. Indeed, with such a narrow view of hinges as rules, the *framework reading* remains blind to the transition hinge-propositions actually undergo: evolving from propositional conviction to non-propositional, non-ratiocinated *action*.[11] Although the *framework reading* falls short, philosophical *therapy*, on the other hand, helps facilitate this transition to fundamental convictions as non-propositional action. Let's consider how.

Appearances to the contrary, Wittgenstein is not being careless. Unlike Moore, he knows he cannot defend our most basic beliefs by decree or proclamation; defending hinge-propositions will require an altogether different approach; something markedly new.[12] Declaring hinge-propositions "exempt from doubt" is one stage in the transition our core convictions undergo.

---

[10] Notice, he is not using the *singular* word, "indeed," there are *two separate* words here, "in" and "deed."

[11] For her part, Coliva does acknowledge that there is "something important to learn from the therapeutic reading" (Coliva 2016: 15).

[12] It is worth pointing out, once again, as the editors of *On Certainty* remind us, this "is all first-draft material, which [Wittgenstein] did not live to excerpt and polish" (Wittgenstein 1972: vi). Avrum Stroll has observed that "[m]any of the entries have the status of first thoughts, something to be put down on paper for further reflection or reconsideration" (Stroll 1994: 80). And, as David Stern points out in his work, *Wittgenstein's Philosophical Investigations: An Introduction*, Wittgenstein's

Despite the obvious problems with the Moorean approach, Wittgenstein still retains Moore's ideas about our convictions, never completely discarding these from his mind. This is evidenced in the fact that Wittgenstein, in an attempt to work through this, reflects on the somewhat disparate beliefs that, on the one hand, our core convictions are something propositional—hinge-propositions, *and*, on the other, that they are "exempt from doubt."[13] Philosophical therapy insists that in order to *reshape negative or inaccurate thinking*, it is imperative that old beliefs never be completely forgotten. We must not "terminate a disease of thought" (Wittgenstein 1967: §382). At §204, in arguably his most important discovery in *On Certainty*, we find Wittgenstein applying this ideal, reshaping what he *had* been calling "hinge-propositions" into non-propositional *action*:

> Giving grounds, however, justifying the evidence, comes to an end;-but the end is not certain propositions' striking us immediately as true, i.e. it is not a kind of seeing on our part; it is our *acting*, which lies at the bottom of the language-game.

As the *propositional* character of the hinge falls away, something *non-propositional* remains in place. At this stage in its evolution, hinges are no longer *propositional*. What stands-fast for us *now*—what frames the background of our thoughts and statements *now*—is located in what we unreflectively *do*. Wittgenstein calls this unreflective action, *certainty*.

Several references to actional certainty can be found throughout the work. For instance:

> §148: Why do I not satisfy myself that I have two feet when I want to get up from a chair? There is no why. I simply don't. This is how I *act*.
>
> §358–9 Now I would like to regard this certainty, not as something akin to hastiness or superficiality, but as a form of life…But that means I want to conceive it as something that lies beyond being justified or unjustified; as it were, as something animal.
>
> §411 If I say "we *assume* that the earth has existed for many years past" (or something similar), then of course it sounds strange that we should *assume* such a thing…The assumption, one might say, forms the basis of action…
>
> §414: What we have here is a foundation for all of my action. But it seems to me that it is wrongly expressed by the words 'I know.'
>
> §475: I want to regard man here as animal; as a primitive being to which one grants instinct but not ratiocination. As a creature in a primitive state.

---

"way of writing and thinking…led him to continually rewrite and rearrange his work." (2004, xiii). I don't wish to make the argument that Wittgenstein simply needed to go back over and polish his thoughts. Rather, from a general point of view, I think we need keep in mind that Wittgenstein, being something of a perfectionist, was quite keen on writing and *re*writing his work; had he the time, perhaps he would have said this differently.

[13] Indeed, he had been interested in "[t]he propositions…which Moore retails as examples of such known truths" for years (Wittgenstein 1972: §137). This interest can be traced back as early as 1933–1934 In the *Blue and Brown Books*, Wittgenstein says: "We seem to have made a discovery—which I could describe by saying that the ground on which we stood and which appeared to be firm and reliable was found to be boggy and unsafe.—That is, this happens when we philosophize; for as soon as we revert to the standpoint of common sense this *general* uncertainty disappears" (Wittgenstein 1958: 45).

And of course, Goethe's quote:

§402: "In the beginning was the deed."

My certainty that the earth exists, that I have two hands, and countless other basic beliefs are borne out, unreflectively, "in the way I *act*" (Wittgenstein 1972: §395). My certainty that the earth exists, for example, is exhibited in actions that I rarely (if ever) think about, e.g., I walk on the earth, I plant trees on it, I swim in its oceans. My certainty that I have two hands is not located in a proof or in an argument, rather, it is demonstrated in what I *do* with my hands, e.g., I greet friends, put on gloves, build a bookshelf, etc. "Sure evidence," says Wittgenstein, "is what we accept as sure, it is evidence that we go by in *acting* surely, *acting* without any doubt" (Wittgenstein 1972: §196). This "acting surely" does not occur at a ratiocinative level, that is, there is nothing to evaluate, contemplate, or weigh in our minds. My actions are the result of the unreflective belief I have when I directly take hold of an item, like a towel.

§510–1: I take hold of my towel without having doubts. And yet this direct taking-hold corresponds to a *sureness*, not to a knowing.

As the "sureness" of our core-beliefs is non-propositional, non-ratiocinated, and actional, it does not conform to traditional epistemic evaluations of truth or justification, this certainty "lie[s] beyond being justified or unjustified" (Wittgenstein 1972: §359).

These passages not only demonstrate Wittgenstein's conviction that certainty is borne out in our unreflective actions, I think these passages also nicely illustrate Wittgenstein's continuing self-therapy, his continuing inner-dialogue leading to deepened understanding.

## 5.5 Writing "with Confidence"

*Philosophical therapy* provides us with the tools necessary in overcoming traditional philosophical problems. Identifying the troubling situation, becoming aware of our thoughts and beliefs, recognizing negative or inaccurate thinking, these are all necessary and important in curing the "philosophical disease." The final stage, however, *reshaping negative or inaccurate thinking*, is perhaps the most challenging of the four stages, as this stage requires the patient hold the toxic thoughts close while simultaneously taking in new ideas. With the old view so close, slipping back into the "familiar routine" is a real and persistent threat. However, if done slowly, methodically, and with care, this final stage also has the potential to be rewarding, yielding beneficial mental well-being for the patient.

From a therapeutic standpoint, an interesting similarity emerges between what Wittgenstein is attempting to do here, in *On Certainty*, and what we find occurring in the Platonic dialogues. In both cases, philosophy is active, social, and dialogical. We see that Wittgenstein needs Moore[14] as much as Plato needs to imagine the likes

[14] And his imaginary interlocutors in *Philosophical Investigations*, too!

of Euthyphro and Crito. The key difference, of course, is that with Plato, the dialogue is *external*, while with Wittgenstein it is *internal*.[15]

As one reads *On Certainty* it becomes clear the extent to which Wittgenstein experienced self-doubt about what he was saying. In passages and parenthetical remarks throughout the work, we find him mocking his own philosophical abilities:

> I do philosophy now like an old woman who is always mislaying something and having to look for it again: now her spectacles, now her keys (Wittgenstein 1972: §532).

He compares his inability to express himself to *Don Quixote*:

> §400: Here I am inclined to fight windmills, because I cannot yet say the thing I really want to say.

He belittles his thinking as "badly thought" (Wittgenstein 1972: §358) and condemns his writing as "badly expressed" (Wittgenstein 1972: §358). From a therapeutic perspective, these notes are sometimes difficult to read. Yet, despite these doubts, Wittgenstein remained confident in what he was *trying* to say. More than half way through the work, in a parenthetical remark, he reflects:

> I believe it might interest a philosopher, one who can think himself, to read my notes. For even if I have hit the mark only rarely, he would recognize what targets I had been ceaselessly aiming at (Wittgenstein 1972: §387).

## References

Benjamin, Martin. 2003. *Philosophy & This Actual World*. Lanham: Rowman & Littlefield Press.

Coliva, Annalisa. 2016. Which Hinge Epistemology? In *Hinge Epistemology*, ed. Coliva and Moyal-Sharrock. Leiden: Brill Publishing.

Goethe, J.W. 1962. (Originally published in 1808). *Faust: Der Tragödie, Part I*. New York: Bantam Dell.

Monk, Ray. 1990. *Ludwig Wittgenstein: The Duty of Genius*. New York: The Free Press.

Moore, G.E. 1925. Defense of Common Sense. In *Philosophical Papers 1962*. New York: Collier Books.

———. 1939. Proof of An External World. In *Philosophical Papers 1962*. New York: Collier Books.

Plato. 1961. *The Collected Dialogues of Plato*. Princeton: Princeton University Press.

Ryle, Gilbert. 1949. *The Concept of Mind*. Chicago: University of Chicago Press.

Stroll, Avrum. 1994. *Moore and Wittgenstein on Certainty*. Oxford: Oxford University Press.

Wittgenstein, Ludwig. 1953. *Philosophical Investigations*. Trans. G.E.M. Anscombe. Oxford: Blackwell Publishers.

———. 1958. *The Blue and Brown Books. Preliminary Studies for the Philosophical Investigations*. Oxford: Basil Blackwell.

———. 1967. *Zettel*. ed. G.E.M. Anscombe and G. H. von Wright , trans. G. E. M. Anscombe, 2nd ed. Oxford: Blackwell Publishers Ltd.

———. 1972. *On Certainty*, ed. G.E.M. Anscombe and G. H. Von Wright, trans. D. Paul and G.E.M. Anscombe. New York: Harper & Row.

———. 1982. *Last Writings On The Philosophy of Psychology*, Vol. I, ed. G.H. von Wright and Heikki Nyman, and trans. C.G. Luckhardt and Maximilian A. E. Aue, Oxford: Blackwell Publishers.

[15] Though, of course, Plato had to be aware of his dialogues internally, as well.

# Chapter 6
# The Language-Game of Knowledge, Hinge-Propositions, & Actional Certitude

**Abstract** This chapter provides an account of the relationship between the "language game" of knowledge, "hinge-propositions," and "actional certainty." This chapter also explores the relationship between two different ways we acquire certainty, characterized as "Bottom-Up" and "Top-Down."

**Keywords** Actional certainty · Arrive at · Begin with · Bottom-up · Common Sense · Hinge-proposition · Instinctual · Knowledge · Language-game · G.E. Moore · Top-down · Unreflective · Ludwig Wittgenstein

## 6.1 The Language-Game of Knowledge

Quite late in *On Certainty*, and with only 10 days remaining before his death, Wittgenstein makes a profound observation: "the concept of knowing," he says, "is coupled with that of the language-game" (Wittgenstein 1972: §560). At first blush, this statement may seem rather unexceptional—so Wittgenstein thinks the concept of knowledge—the claims of, or disputes about, the nature and boundaries of knowledge—is part of a language-game, the language-game of knowledge…Why is *this* significant?

In order to understand just how important it is, recall the central position that epistemology has traditionally occupied in Western philosophy. At least since the time of Descartes, the nature, sources, and limits of knowledge were thought to be a main pillar of the discipline. Indeed, it is still defined in some encyclopedias as one of the discipline's "core areas."[1] But by describing "the concept of knowledge" as an element of a *language-game*, Wittgenstein is effectively stripping it of its central importance. Epistemology, he tacitly implies, is on par with countless *other*

[1] *Concise Routledge Encyclopedia of Philosophy*, 2000. New York: Routledge. s.v. "epistemology." p. 246.



R. G. Brice, *Wittgenstein's On Certainty: Insight and Method*, SpringerBriefs in Philosophy, https://doi.org/10.1007/978-3-030-90781-5_6

language-games,[2] thus denying it its privileged position. If claims or debates about knowledge are nothing but moves within yet another language-game, then knowledge is not more basic or fundamental than other areas of philosophy, for it too must satisfy rules that are established within *its* discipline, within *its* language-game. This might include responding to questions like: is this proposition *true*? Is this proposition *justified*? Claims within the language-game of knowledge must attempt to demonstrate that rules have been followed; such claims must attempt to demonstrate that questions posed have been responded to.

Implicitly, Wittgenstein's observation here illustrates an important division: actional certainty is categorically *distinct* from knowledge. Certainty takes on a foundational role: it is expressed in our hinge-*propositions* which, in turn, supports our language-games—*including* the language-game of knowledge.

## 6.2 "Different Categories"

With Wittgenstein's careful guidance, in *On Certainty* we come to see that our most basic convictions, *certainties*, must be understood as both unreflective *and* non-propositional, divorced from any and all *language-games* and removed from *hinge-propositions* as well. This is where, in the end, Wittgenstein locates certainty. *And*, because this certainty is *sui generis*, a category of its own, it differs from the way philosophers have traditionally understood it, as a *degree* of knowledge. As illustrated in Chap. 1, from Aquinas to Russell, philosophers have typically expressed certainty as a *level* within the body of knowledge, employing some variation of the phrase, "I know *with certainty* that *p*." What Wittgenstein's work in *On Certainty* shows us is that knowledge and certainty do not differ in degree, but rather, in *kind*—they "belong to different *categories*" (Wittgenstein 1972: §308).

The first indication of a categorical difference between knowledge and certainty emerges early in *On Certainty*. In §8 he says:

> The difference between the concept of 'knowing' and the concept of 'being certain' isn't of any great importance at all, except where 'I know' is meant to mean 'I *can't* be wrong' (Wittgenstein 1972: §8).

This is not simply a linguistic distinction. Wittgenstein is drawing out—identifying—the categorical divide here between these two concepts. And while his implicit point about philosophers using the terms "knowledge" and "certainty" haphazardly, even *interchangeably*, may be true (perhaps even contributing to some extent to the "philosophical disease" he hopes to cure), his main point is that these two concepts

[2] Indeed, besides the language-game of knowledge, there are also mathematical language-games, logical language-games, language-games of aesthetics, of color, of metaphysics, of religion; and countless others.

are categorically *distinct*.[3] He returns to this distinction in many passages throughout *On Certainty*. For example, §94:

> ...I did not get my picture of the world by satisfying myself of its correctness; nor do I have it because I am satisfied of its correctness. No: *it is the inherited background against which I distinguish between true and false*.[4]

Some things, many things, are in place ("inherited") right from the start. We rarely talk or even think about them. Although they go unnoticed, at some non-cognitive level they *are* implicit in our minds. For instance, if I ask a 3 year-old to put her shoes on, her inherited and unreflective conviction will include, among other things, that her shoes, socks, and feet, all occupy space. This unreflective conviction contributes to her many other inherited foundational beliefs, which, in turn, allow her to make sense of her world. They form a system; and when contexts require these beliefs move from tacit to *reflective* conviction, they form a "system of propositions."

> §141 When we first begin to *believe* anything, what we believe is not a single proposition, it is a whole system of propositions. (Light dawns gradually over the whole.)
> §142 It is not single axioms that strike me as obvious, it is a system in which consequences and premises give one another *mutual* support.
> §225 What I hold fast to is not *one* proposition but a nest of propositions.

This "nest of propositions" represents our most basic, fundamental convictions, *certainties*.[5]

[3] But even if these terms *were* interchangeable, it still would make no difference to the skeptic who would undoubtedly ask, "Why are you so *certain*?" For the skeptic, it matters little whether you express this in a traditional knowledge claim (e.g., "I know *that p*"), a knowledge claim that indicates a degree of support (e.g., "I know with certainty *that p*"), or even whether you give up framing it with knowledge altogether and instead claim certainty (e.g., "I am certain *that p*."). The issue turns on the skeptic's demand for *grounds*, (e.g., "What grounds have you to make such a claim?"). But as Wittgenstein slowly realizes throughout *On Certainty*, our grounds for "certitude" are *different* than our grounds for "knowing." As we've seen, and as Wittgenstein explains a few passages later, knowledge claims *can* be wrong ("One always forgets the expression 'I thought I knew'" (Wittgenstein 1972: §12), but his purpose here is simply to get us to see that these two concepts are not separated by degree, but categorically divorced.

[4] My emphasis. It is also worth noting that here, in passage §94 he says his "picture of the world" did not have to do with its "correctness." His choice of language is intentional. Indeed, "picture of the world" and its "correctness" are references to his early view in *Tractatus Logico-Philosophicus*. Back then, Wittgenstein believed there was a one-to-one (or isomorphic) relationship between language ("the logical picture of the world") and the world itself. If the language (propositions) used accurately described the state of affairs in the world, then the proposition was deemed "correct" and we would simply supply it with a truth-value of "true." The later Wittgenstein, however, rejected this "picture of the world" in favor of meaning, as understood through our *language-games* and *forms of life*.

[5] An important but underdeveloped consideration that exists in *On Certainty* is how we go about *acquiring* this "nest." We will consider this important aspect of certainty acquisition in more detail below, in Sect. 6.5.

In another series of passages (§§148, 150–152), Wittgenstein counters Moore's reflective knowledge claim that he has two hands with his *non*-reflective (or reflexive), actional *certainty* that he has two feet:

> §148 Why do I not satisfy myself that I have two feet when I want to get up from a chair? There is no why. I simply don't. *This is how I act.*

Wittgenstein's point here is this: I don't look to see if my feet are still attached to my legs when I want to get up from a chair—I'm certain of it and I just *do* it. If, for some reason I am *not* certain, it is not clear why looking will provide me with the necessary justification. This is the case whether I have two feet or two hands. As he says,

> If a blind man were to ask me, "Have you got two hands?" I should not make sure by looking. If I were to have any doubt of it, then I don't know why I should trust my eyes. For why shouldn't I test my *eyes* by looking to find out whether I see my two hands? (Wittgenstein 1972: §125).

Indeed, if I had difficulty trusting whether or not I had two hands, it seems my lack of trust would almost certainly have to extend to my eyes as well. I don't think about my having feet…I simply get up. I don't think about my having *hands*, I simply use them to open doors, tie my shoes, wave to my friends, etc. *This*, to answer Wittgenstein's question in §148, is how skepticism is to be "satisfied," or better, *this* is how radical skepticism is to be kept in check. Supplying the skeptic with reasons only encourages him to remain within his familiar, but misleading, paper-and-pencil framework—a framework in which everything relevant or important is, or can be, stated or written down in words. This sort of limitation can make one feel trapped or claustrophobic. Therapeutic philosophy's goal throughout all of this must be to show him a way *out*. As Wittgenstein puts it elsewhere, the "aim in philosophy…[is] to show the fly the way out of the fly-bottle" (Wittgenstein 1953: §309). Returning to the other passages in this series:

> §150 How does someone judge which is his right and which his left hand? How do I know that my judgment will agree with someone else's? How do I know that this color is blue? If I don't trust *myself* here, why should I trust anyone else's judgment? Is there a why? Must I not begin to trust somewhere? That is to say: *somewhere I must begin with not-doubting*; and that is not, so to speak, hasty but excusable: it is part of judging.[6]
>
> §151 I should like to say: Moore does not *know* what he asserts he knows, but it stands fast for him, as also for me; regarding it as absolutely solid is part of our *method* of doubt and enquiry.
>
> §152 I do not explicitly learn the propositions that stand fast for me. I can *discover* them subsequently like the axis around which a body rotates. This axis is not fixed in the sense that anything holds it fast, but the movement around it determines its immobility.[7]

[6] My emphasis.

[7] Similarly, at §144 Wittgenstein says: "The child learns to believe a host of things. I.e. it learns to act according to these beliefs. Bit by bit there forms a system of what is believed, and in that system some things stand unshakably fast and some are more or less liable to shift. What stands fast

The metaphor in this last passage nicely captures the non-reflective acceptance of our basic convictions. It is true, we are hardly ever *aware* of the earth's axis, but at some level we are mindful that it exists; for earth, and everything on it, revolve around it. Our basic propositions are like this. The unreflective action/s that we perform around them demonstrate how they stand fast for us. Indeed, many propositions we learn depend on other, more fundamental ones, but because these more fundamental propositions are so basic, we rarely consider them—like the earth's axis, we are not consciously aware of them. When they *are* exposed, they seem banal, even absurd.[8]

## 6.3 Hinge-Propositions: A Turning Point

In Chap. 5, we discussed the essential role hinge-propositions play in our lives. These core, Common Sense convictions anchor our language, our practices, and our activities. Because so much else turns on them they "must," as Wittgenstein tells us, "stay put" (Wittgenstein 1972: §343). He also asserts that hinge-propositions are "exempt from doubt" (Wittgenstein 1972: §341). This, we said at the time, seemed suspicious. By insisting they were *not* subject to doubt, Wittgenstein appeared to be committing Moore's mistake all over again, namely, that there are some things we just cannot question. Because these convictions are expressed *as propositions*, Wittgenstein seemed to be *inviting* the skeptic to ask: "But how do you know?"

When viewed through our therapeutic lens, we located this problem in the final and most challenging stage of our process—*reshaping negative or inaccurate thinking*. At this stage, the patient must preserve two (sometimes *conflicting*) thoughts in his mind before attempting to reshape them into something new. He must hold the old view close enough to analyze it, but at the same time remain distant enough to avoid being "pulled back in." If done cautiously, and with the assistance of a therapist (or, like Wittgenstein, a self-therapist), this stage can serve as the final step in fostering a new way of thinking about philosophical problems. By encouraging philosophers to reshape old patterns of thought, this process can help form a whole new philosophical life.

For Wittgenstein, this seemed to mark a turning point. Recall, at this stage, he effectively reshaped these conflicting thoughts into something new: he showed that hinge-propositions are *expressed as*, but are not *actually*, propositions. And once the

---

does so, not because it is intrinsically obvious or convincing; it is rather held fast by what lies around it." This "system" forms our hinge-propositions (we rarely reflect on these beliefs).

[8] The same idea might be expressed in the context of a pedestrian avoiding an on-coming car. After successfully avoiding the car, assume someone asks him the ridiculous question, "Why did you move?" How might he respond? Surely something like, "Well, I didn't want to get hit!" If pressed, however (and although it would sound absurd), he might say, look "I got out of the way because I have a body, and having a body, means you make an effort to *preserve it*!" This proposition stands fast for him, as it does for you and for me, but we rarely think about such basic propositions.

propositional character is gone, these hinges thus *become* actional certitude. Wittgenstein's breakthrough here was realizing that what he and Moore had been *trying* to say *could not be said*. Our basic convictions are borne out in the things we unreflectively *do*, not in the things we reflectively *say*. This proved to be significant, not only for Wittgenstein personally, but also, on a much larger scale, for the discipline of philosophy, and for philosophers infected with the "disease of thought"—a disease that assumes everything of philosophical importance is fundamentally linguistic, expressed in words, by pencil on paper.

## 6.4 The Interrelation of the Language-Game of Knowledge, Hinge-Propositions, & Actional Certainty

Another issue discussed in Chap. 5 was the fact that knowledge claims consist of beliefs (opinions) that are not only *true* but *justified*.[9] This remains the case—even though we now locate those claims to knowledge in a language-game. Indeed, we might now describe tests for truth and justification as some of the rules that apply to the language-game of knowledge. Regardless where knowledge is now situated, it nonetheless retains a problem. Although a particular knowledge claim may succeed *now*, at this particular time and in this particular place, it does not mean it will *always* remain that way. Knowledge claims about the world are not absolute. In other words, what we know *now* may, in time, need to be modified, revised—even abandoned in light of changes in our new understanding of the world. The question facing epistemologists is whether this means that knowledge is, in a pejorative sense, *merely* relative.[10]

This is an old philosophical debate. In the past, philosophers have addressed it by making a distinction between *absolute* and *contingent* truths (as well as *deductive* versus *inductive* reasoning). In order to see how Wittgenstein addresses this question, let's turn to a set of passages where relative or contingent propositions are prominently discussed.

[9] Much has been written on *justified true belief* since Edmund Gettier's influential 1963 paper, "Is Justified True Belief Knowledge?" This paper sparked a debate that is quite broad and presents difficult interpretive and philosophic problems that are not considered here.

[10] W.V.O. Quine, in "Two Dogmas of Empiricism," says, "no statement is immune to revision" (Quine 1980: 43). However, some 20 years after the publication of "Two Dogmas," he addressed this question in the following way: "[s]uppose that from a combined dozen of our theoretical beliefs a scientist derives a prediction in molecular biology, and the prediction fails. He is apt to scrutinize for possible revision only the half dozen beliefs that belonged to molecular biology rather than tamper with the more general half dozen having to do with logic and arithmetic and the gross behavior of bodies. This is a reasonable strategy—*a maxim of minimum mutilation*" (Quine 1986: 7, my emphasis). Although Wittgenstein offers no such *maxim*, he seems to be developing a similar strategy.

Interspersed throughout *On Certainty* are some 16 or so passages where Wittgenstein offers some variation of the phrase, "No one has been to the moon." Let's consider just two of these "moon passages," §106 and §108.[11]

> §106 Suppose some adult had told a child that he had been on the moon. The child tells me the story, and I say it was only a joke, the man hadn't been on the moon; no one has ever been on the moon; the moon is a long way off and it is impossible to climb up there or fly there.—If now the child insists, saying perhaps there is a way of getting there which I don't know, etc. what reply could I make to him?...But a child will not ordinarily stick to such a belief and will soon be convinced by what we tell him seriously.
>
> §108 "But is there...no objective truth? Isn't it true, or false, that someone has been on the moon?" If we are thinking within our system, then it is certain that no one has ever been on the moon. Not merely is nothing of the sort ever seriously reported to us by reasonable people, but our whole system of physics forbids us to believe it. For this demands answers to the questions "How did he overcome the force of gravity?" "How could he live without an atmosphere?" and a thousand others which could not be answered. But suppose that instead of all these answers we met the reply: "We don't know how one gets to the moon, but those who get there know at once that they are there; and even you can't explain everything." We should feel ourselves intellectually very distant from someone who said this.

Eighteen years after Wittgenstein's death, the proposition, "no one has been to the moon," was no longer true. In fact, the belief that it is "impossible to...fly there" became demonstrably false. The "thinking within our system" had changed. NASA physicists, scientists, and engineers contemplated these questions and succeeded in developing answers that were previously *outside* of our system. Had Wittgenstein lived, his demand for answers to the questions, "'How did he overcome the force of gravity?' 'How could he live without an atmosphere?' and a thousand others," would have been met.

Despite these issues, Wittgenstein's "moon passages" should not be considered wrong. On the contrary, these passages nicely illustrate a *relative* certainty—what I have elsewhere referred to as "particular hinge-propositions" (Brice 2014: 36, 49, 57, 97). A particular hinge-proposition serves as a "provisionally fixed point" (Rawls 1971: 19), it occurs *within* a language-game. Particular hinge-propositions are based on the best available information and data we have at the time yet, they remain vulnerable (relative) to *newer* information and *better* data. In 1951, the data informed us that it was not possible to get to the moon, as Wittgenstein rightly points out; but as improved equipment led to correspondingly improved information and data, eventually it was determined that we *could* in fact get to the moon, hence, the original proposition was overturned.

Hinge-propositions internal to particular language-games are not explicitly discussed in *On Certainty*, but their presence is unmistakable. These relatively certain propositions often remain *within* a language-game, however, they sometimes work their way out to *become* an actional certainty. But this raises the question: how? *How* do they *become* actional?

---

[11]Aside from the two mentioned above, the remaining passages include: §§111, 117, 171, 226, 238, 264, 269, 286, 327, 332, 337, 338, 661, and 662.

As a point of entry here, consider how some of our social and cultural influences ultimately result in non-reflective behavior. For example, take the unreflective act of putting on a seatbelt, using a fork and knife, holding the door open for someone immediately behind you. Each of these acts no longer requires thought, after some time and after having received positive reinforcement from family and others, the action is performed reflexively. This delivers us to two very different ways in which we go about acquiring our basic beliefs.

Sometimes, in *On Certainty*, Wittgenstein discusses convictions we accept without question—those we unreflectively and instinctively *begin with*. Other times, however, he discusses certainty as something we *arrive at*—convictions we develop through scientific investigation, through social/cultural conditioning, etc. These two ways of acquiring certainty are important from a philosophical perspective, but they also fit nicely with Wittgenstein's therapeutic approach—offering a scope and limit on how far skepticism can be pushed, and a signal to the therapist and, more importantly, the patient, that Common Sense tools must be implemented now to prevent the philosopher from reverting to fruitless forms of inquiry.[12]

## 6.5 Tracking Its Acquisition: Bottom-Up & Top-Down Certainty

In tracking certainty's path, two trajectories quickly emerge: sometimes our unreflective, non-propositional convictions are ones we cannot help *beginning with*—these travel upward (*Bottom-Up Certitude*); others are convictions that we *develop*, as a culture and as a human being participating in that culture—these convictions travel downward (*Top-Down Certitude*). We can briefly define these two trajectories in the following way:

1. *Bottom-Up*: the most obvious characteristic of the bottom-up model is the fact that basic, fundamental convictions are something we unreflectively *begin with*; they are typically instinctual. If/when we reflect on these convictions, they travel *up*, so to speak, to our basic propositional level, to *become* hinge-propositions.
2. *Top-Down*: on this model, through training and positive reinforcement by family and culture, we *arrive at* unreflective, actional certainties. Beginning in a language-game, as we repeatedly perform an action, it travels down, so to speak, slowly *becoming* habitual; in time, the conviction *becomes* second nature.

In the bottom-up model, our non-reflective, actional certitude plays a foundational role, not just for our hinge-propositions, but for our entire network of interrelated and overlapping language-games, too. In the top-down model, however, actional certitude *begins* as something unclear and less defined. As it is slowly shaped by

[12] I've written about these two ways of acquiring certainty elsewhere. For a fuller account, see Brice 2014.

customs and attitudes within our individual language-games and corresponding *form of life*, it *becomes* unreflective—ultimately demonstrated *in our actions*. Philosophical therapy plays an important role here, providing oversight and guidance by reinforcing the scope and limit of fruitful inquiry. Philosophical therapy can help prevent us from "going too far" by guiding us and directing us where to install our Common Sense guardrails.

### 6.5.1 Bottom-Up Certitude

In the bottom-up model, certainties are convictions we unreflectively *begin with*—they constitute our instinctual beliefs; these certainties are non-reflective, actional, and foundational. Although we don't ordinarily state or reflect on them, if/when we *do*, these convictions travel *up* to our basic propositional level, they *become* hinge-propositions. Let's consider how this works.

While the study of evolutionary biology may be able to tell me how it has come about that I protect my child from an oncoming car—perhaps a neurophysiologist can even pin-point what region of the brain is active when I push her out of the way—my *understanding* of this occurs *later*, at a higher level of ratiocination. When *this* occurs, the previously unreflective conviction has travelled *up* to a reflective level of thought, before travelling *higher still*, to a point within a language-game. (Maybe the language-game of evolutionary biology, or the language-game of neurophysiology.)

It is important not to confuse or replace the instinctual *cause* here with a *reason* or a *justification*. As Wittgenstein warns us, we should not be asking the question: "'What goes on in us when we are certain that…?'—but: How is 'the certainty that this is the case' manifested in human action?" (Wittgenstein 1953: II xi, p. 225). In other words, how is our certainty *borne out in the action*? There is no need for reason or justification at the *actional* level. Maybe in hindsight, and at a higher, more reflective level we might offer a reason, but at the time it occurs, our reply need only be: "This is simply what I *do*" (Wittgenstein 1953: §217).

From a therapeutic perspective, this provides us with a necessary, and comforting, backstop. We need not answer extraordinary questions formulated by the skeptic, for our unreflective actions serve to demonstrate what our unstated convictions are here. But even if/when we *do* respond, Wittgenstein, in his role as therapist here, says we need only reply: "This is what I do."

As we have seen, hinge-propositions form a special class; they represent in words, the beliefs or convictions that are almost always accepted, without thought, in the things we do. Hinges are, in effect, *propositionalized* certainty; that is, selected, nonverbal, fundamental convictions expressed or conveyed as "propositions." This propositionalized certainty plays a foundational role for our many different language-games. But at the hinge-proposition's base—*its* foundation—is non-propositional, non-reflective, *actional* certitude. This represented an important turning point for Wittgenstein, a point where he realized hinge-propositions, when shed of their propositional character, represent our actional certainties.

Hinge-propositions serve as an important in-between. For in the bottom-up model, while they are themselves foundational—supporting our entire network of different language-games—they are also propositional and ratiocinative and, as we've seen, this means they must allow skeptical questions, regardless how extraordinary, absurd, or bizarre.[13] While "propositionalized" actional certainties are verbal and may invite skeptical questions, the answers to those questions are not further or additional propositions, but rather acting or pointing to what we in fact *do*.

Moreover, note how we readily diagnose people (non-philosophers) who carry on as if the earth does not exist or as if *their* mental states are the *only* mental states there are; they are diagnosed as suffering from a "mental disturbance." We do *not* extend this diagnosis to those who refuse to respond to their bizarre questions. Bottom-up certainty helps us contain the philosophical disease from spreading. It demonstrates where the line between acceptable and unacceptable skeptical questions must be drawn. As such, it helps contain the philosophical disease of Cartesian doubt.

While this might explain our acquisition of the kinds of certitudinal belief we *begin with*, it would be wrong to assume it is limited to just this one way (*Bottom-Up*). Consider, for instance, the hinge-proposition, "Water boils at 100° C" (Wittgenstein 1972: §567). This conviction is *learned*, and once accepted, it is no longer something we doubt. Convictions like this serve to highlight the fact that, as D. Z. Phillips, points out, "Wittgenstein is not aiming for a consistent account of all the [hinge] propositions he discusses" (Phillips 2003: 154). That is to say, he is not trying to classify *all* hinge propositions as foundational. Indeed, "in some cases,"

> what is *now* accepted without question was *arrived at* as the result of an investigation, for example, that every skull contains a brain. What is true is that once discovered…no one would doubt that every skull contains a brain (Phillips 2003: 155).

Certainty is *not*, in all cases, a matter of biological instinct—something we *begin with*. As Phillips correctly points out, in many cases, certainty is something we ultimately *acquire*, either through education, scientific discovery, or the positive reinforcement we receive from family and culture. Our basic convictions, our "nest of propositions," can develop through what we have been taught—the things we've learned. Over time, within particular language-games, these propositions can lose their propositional character and drop down to a non-reflective, *actional* level. There is a sense in which, while having their origins in particular language-games, top-down certainties often *become* second nature.[14]

---

[13] This is where Coliva and Moyal-Sharrock go wrong in their *framework approach*.

[14] Throughout the years, philosophers have used the phrase *second nature* to describe a kind of certainty acquisition. Jose Medina uses it to describe a sort of naturalism found in the *third Wittgenstein*; John McDowell uses it in several places throughout his book, *Mind and Will* (McDowell 1996: 84–88, 91–92, 103–104, 109–100). Most notably, however, the phrase can be traced back to Aristotle who, in *Nicomachean Ethics* 1152a30 uses it while discussing the difficulty that exists when we try to alter the sorts of actions that have become habit in us: "habit is hard to change," he says, and that's "because habit is a sort of *second nature*" (Aristotle 2013: 1152a30). To be fair, there has been *some* disagreement on this translation. W.D. Ross, for example, translates it as "because it is *like nature*" (Ross, W.D. 1984. *The Complete Works of Aristotle* Princeton: Princeton University Press), while H. Rackham translates it as "because it is a *sort of nature*" (Rackman, H 1934. *Aristotle Nicomachean Ethics* Cambridge: Harvard University Press).

### 6.5.2 Top-Down Certitude

If we were to describe *second nature* in Wittgensteinian terms, we might think of it as one that has its start in a language-game before dropping down to the level of non-reflective, non-propositional actional certitude. There comes a point where we no longer contemplate *what* we are doing—we simply *do it*. In the top-down model, the process begins with propositions in a language-game. With positive reinforcement from family, teachers, etc., these propositions become less and less cognitive; in time, "arriving at" a level demonstrated in unreflective behavior—*actional certitude*. Consider a few of the propositions Wittgenstein uses in *On Certainty*:

- "Water boils at 100 °C" (Wittgenstein 1972: §567).
- "I am in England" (Wittgenstein 1972: §421).
- "I am sitting writing at the table" (Wittgenstein 1972: §675).

While these propositions represent some rather basic convictions, they initially are the result of teaching and of positive reinforcement. They are the product of having lived in a particular culture, a particular *form of life* and, at least in the beginning, they are not *automatically* accepted. Our comprehension of concepts like measuring temperature, geographic location, even what it means to both be seated and write comprehensible sentences, only make sense (only take on meaning) *within* a language-game within a culture, so the source of these convictions cannot be instinctive. In time, they often *become* actionally certain, but this is due to the repetition that occurs early on in our *form of life*.[15] Because these responses have been drilled into us from the start, we ultimately arrive at them without thought and "in a flash" (Wittgenstein 1953: §§138–139). For instance, without thought, a Syrian man, before entering a mosque, removes his shoes; without thought, an American businesswoman, after meeting a new client, extends her hand for a handshake; without thought, a Japanese man bows his head to thank a stranger for retrieving his lost wallet. These are all instances of what we might call, "top-down certainty." The actions *become* something performed without thought. Indeed, most of us become proficient in our response to a given rule or custom in our culture/society, but it is through training and positive reinforcement that we develop this kind of *second nature*.[16]

[15] Second nature seems to also be the focus in §568. "If one of my names were used only very rarely, then it might happen that I did not know it. It goes without saying that I know my name, only because, like anyone else, I use it over and over again." It is true you use it over and over again—in different settings, addressing different audiences, at different times, but the point here is that first, you come to know it, and *then* it becomes second nature.

[16] A rather interesting connection exists between this kind of certainty acquisition and something we discussed back in Chap. 2 on the origins of CBT and its connection to Common Sense. Recall, while tracing the roots of CBT, we discussed the contributions of the Scottish surgeon, James Braid, who helped reunite philosophical methods with therapeutic strategies through a careful analysis of *hypnosis*. Hypnosis proved less mysterious than initially thought, as Braid successfully induced the trance-like state in subjects using methods he developed from his association with

In both the bottom-up and top-down models, Common Sense limitations are firmly in place. Common Sense boundaries prevent us from questioning whether there are other minds or questioning whether the earth exists. Common Sense prevents us from questioning the temperature of boiling water, one's geo-location, etc. From a therapeutic perspective then, when propositions in the top-down model drop to an unreflective level, actional certitude serves to prevent further inquiry. For if someone determinedly failed to, say, drive on the right side of the road (at least in the United States), we might find this action an indicator of "mental disturbance."

### 6.5.3 *Differences*

A big difference between the bottom-up and the top-down models is that the guardrails in the latter are determined socially/culturally, within some language-game within a related *form of life*; in the former, however, our Common Sense limitations, our guardrails, are determined through instinct—here it is, as Wittgenstein puts it, "animal" (Wittgenstein 1972: §475).

Because the guardrails in the two models are determined differently, another important difference finds that these propositions, these hinge-propositions, are not basic to *all* language-games. Within the top-down model, learned propositions, although basic to a *particular* language-game, are not basic to *all* of them and therefore should not be confused with the hinge-propositions we find in the bottom-up model. In other words, the top-down model's certainty is acquired differently. We "arrive at" or acquire this certitude, *pace* Phillips, we do not *begin* with it. These propositions are not hinge-propositions, *per se*, rather, they are, what I have

---

Thomas Reid's Scottish School of Common Sense. Hypnosis, he concluded, could serve as a kind of therapeutic curative. As such, Braid played a vital role in linking the early stages of CBT with Common Sense. Working closely with Braid was the distinguished neuropsychologist, William Benjamin Carpenter. As Robertson explains in *The Discovery of Hypnosis*, "...many Victorian empiricists were keen to disprove "occult" theories such as Mesmerism, and to offer more credible scientific explanations in their stead. Most notable among these was Prof. William B. Carpenter, the distinguished physiologist" (Robertson 2013: 18). Carpenter also subscribed to the tenets of Common Sense and, in correspondence with the utilitarian, John Stuart Mill, expressed his belief that Common Sense was demonstrated through "a combination of innate and acquired judgments" (Robertson 2013: 18). These judgments, Carpenter reasoned, "have a reflexive or automatic quality and appear to consciousness as self-evident truths" (Robertson 2013: 18). Mill agreed. "I have long recognized as a fact that judgments really grounded on a long succession have small experiences mostly forgotten, or perhaps never brought out into very distinct consciousness, often grow into the likeness of intuitive perceptions...When states of mind and no respect innate or instinctive have been frequently repeated, the mind acquires, as is proved by the power of habit, a greatly increased facility of passing into those States and this increased facility must be owing to some change of a physical character, in the organic action of the brain" (Mill 1971: 1868). Carpenter and Mill both supported the idea that some of our "rational judgments" evolve into "reflexive judgments" through "the law of habit and association" (Robertson 2013: 18), two methods closely connected with the School of Common Sense. These habituated judgments, in turn, *become* "second nature"—a kind of *learned instinct* (Robertson 2013: 18).

elsewhere called, "*particular* hinge-propositions," particular to specific language-games, e.g., the language-game of morality, aesthetics, religion, etc. (Brice 2014: 36, 49, 57, 97). Unlike hinge-propositions we find in the bottom-up approach, acceptance of *these* propositions requires socialization (enculturation) within a *form of life*. These beliefs (and the others that surround them) are then reinforced in different circumstances, at different times, in front of different audiences within those particular *forms of life*. This is an important difference. Bottom-up certainties are borne out in our actions; they are trivial and obvious to *all* of us, in *any and all forms of life*, e.g., "avoid fast-moving, harmful objects."[17] With top-down certainty, on the other hand, at least some of the actions that *become* unreflective in this model are neither trivial nor obvious to all of us when we utter them. For example, there is nothing trivial nor obvious to us all when a subway operator says, "Switch A must be off when approaching the platform." Although this may seem trivial to people who reside *in* that *form of life*, to the rest of us *non*-subway operators, it is not. These propositions, like countless others, have their *origins in*, and are *governed by*, the rules and conventions of language-games within that culture. Therefore, when we propositionalize the action in the *top-down* model, it *cannot* be located among the most fundamental hinge-propositions; it must reside at the lowest level within a *particular* language-game.

In a particular language-game, we might explain, offer reasons, even justify *why* we do *what* we do. But as we dig deeper, as we burrow further and further down to our most fundamental beliefs, we must resist the urge to *explain*, and instead, as Wittgenstein says, just *describe*.

> §189 At some point one has to pass from explanation to mere description.
> §192 To be sure there is justification; but justification comes to an end.

To offer explanation is to detail *why* we do this or do that, it is an account, a justification of the rules we follow. This works well *within* a language-game, that is, it works well if we are only thinking of hinges as particular *to a language-game*, but there is no level of hinge-propositions *within* a language-game in the top-down model. Once the proposition has been learned, this particular hinge-proposition has worked its way out to become an actional, unreflective certitude. In short, these propositions, while basic to, and found within, *particular* language-games, they are not basic to, nor are they found within, *all* language-games.

For this reason, as discussed in Chap. 4, the *Framework Approach*, promoted by Moyal-Sharrock and Coliva, works fine if we restrict it to the top-down model, but doesn't work in the bottom-up approach. Recall, Moyal-Sharrock and Coliva's argument turned on the idea that hinge-propositions are rules, and rules, they argued, "are required to collect evidence for or against ordinary empirical propositions" (Coliva 2016: 17). They are, Coliva says, "rationally mandated" (Coliva 2016: 17). But, as discussed here and in Chap. 4, it is simply wrong to think of hinge-propositions as rules in the bottom-up model. This is where the *Framework Reading*

[17] I include "harmful" here because, in baseball, apart from hitters avoiding bean balls, fielders do not avoid, but try to catch or field fast moving balls.

goes quite wrong. It either ignores or overlooks the transition hinge-propositions actually undergo when moving, or evolving, from propositional conviction to non-propositional, non-ratiocinated *action*.[18] We can, however, accommodate the *Framework Approach* if we limit it to *only particular hinge-propositions* in the top-down model. The idea here is to recognize that some propositions may *become* certain, passing from a propositional, ratiocinated, rule-based *language-game*, down to—*and becoming*—an unreflective, non-propositional actional certainty. It "arrives at" this actional certitude through positive reinforcement of the rules. Although the *Framework Approach* falters here, *Philosophical therapy*, as we've seen helps facilitate the transition to fundamental convictions as non-propositional action.

In this chapter, we've considered two ways in which we acquire certainty. The bottom-up and top-down methods provide important guardrails to prevent us from "going over the edge" down a dangerous slope. Wittgenstein's therapeutic method, if given a chance, can effectively treat philosophical confusions developing out of the Cartesian method of systematic doubt.

Wittgenstein's psychotherapeutic tools bear a striking resemblance to those found in CBT, however, in philosophy, it is fair to say it was unlike anything anyone had ever seen before. Indeed, near the end of his life, G.E. Moore, reflecting on Wittgenstein's philosophical approach, described it as a "quite different…method, [one] which…I have never been able to understand clearly enough to use" (Moore 1942: 33).

## References

Aristotle. 2013. *Nicomachean Ethics*. Trans. F.H. Peters. New York: Barnes & Noble Publishing.

Brice, Robert Greenleaf. 2014. *Exploring Certainty: Wittgenstein and Wide Fields of Thought*. Lanham: Lexington Books.

Coliva, Annalisa. 2016. Which Hinge Epistemology? In *Hinge Epistemology*, ed. Coliva and Moyal-Sharrock. Leiden: Brill Publishing.

Gettier, Edmund L. 1963. Is Justified True Belief Knowledge? *Analysis* 23 (6): 121–123.

McDowell, John. 1996. *Mind and Will*. Cambridge: Harvard University Press.

Mill, John Stuart. 1971. *The Later Letters of John Stuart Mill 1849–1873: Volumes XIV-XVII*. Toronto: University of Toronto Press.

Moore, G.E. 1942. An Autobiography. In *The Philosophy of G.E. Moore*. La Salle: Open Court Publishing Company.

Phillips, D.Z. 2003. Afterword. In *Rush Rhees Wittgenstein's On Certainty: There—Like Our Life*. Malden: Blackwell Publishing.

Quine, W.V.O. 1980. *Two Dogmas of Empiricism in From A Logical Point of View*. Cambridge: Harvard University Press.

———. 1986. *Philosophy of Logic*. 2nd ed. Cambridge: Harvard University Press.

Rawls, John. 1971. *A Theory of Justice*. Cambridge: Harvard University Press.

[18] As we noted earlier, for her part, Coliva does acknowledge that there is "something important to learn from the therapeutic reading" (Coliva 2016: 15).

Robertson, Donald. 2013. *The Discovery of Hypnosis: The Complete Writings of James Braid, the Father of Hypnotherapy*. Lulu.com

Wittgenstein, Ludwig. 1953. *Philosophical Investigations*. Trans. G.E.M. Anscombe. Oxford: Blackwell Publishers.

———. 1972. *On Certainty*, ed. G.E.M. Anscombe and G. H. Von Wright, and trans. D. Paul and G.E.M. Anscombe. New York: Harper & Row.

# Chapter 7
# Therapeutic Philosophy: "A Quite Different Method"

**Abstract** This chapter focuses on philosophical therapy as a method, providing an account of what, generally, Wittgenstein's method can do for us, and what needs to be done in order for it to happen.

**Keywords** Combative therapy · Common sense · Rene Descartes · G.E. Moore · Peace · philosophical disease · Philosophical therapy · Skepticism · Ludwig Wittgenstein

## 7.1 Running "Its Natural Course"

In 1942, *The Library of Living Philosophers* series published a volume on G.E. Moore.[1] In addition to the philosophical essays and the questions posed to Moore, the volume also contained a brief autobiography. In the few paragraphs he reserves for Wittgenstein, Moore expressed admiration for his ability as a philosopher: "he was much cleverer at philosophy than I was," Moore confessed. He was "much more profound… with much better insight" (Moore 1942: 33). He also thought it important to highlight Wittgenstein's unique method. Moore wrote:

> He has made me think that what is required for the solution of philosophical problems which baffle me, is a method quite different from any which I have ever used—a method which he himself uses successfully, but which I have never been able to understand clearly enough to use it myself (Moore 1942: 33).

[1] From *The Library of Living Philosophers* website: "*The Library of Living Philosophers* (*LLP*), which commenced in 1939 with a volume on John Dewey, is an unparalleled series that has made an advancement to the understanding of philosophy through rational debate. In each volume, a great philosopher presents his views in an intellectual biography. This is followed by a number of essays by distinguished scholars who critique the great philosopher's ideas. The LLP's subject then replies to each critic. Far from being a mere commentary on the life's work of an important thinker, an LLP volume is itself a key part of that life's work." See: http://www.opencourtbooks.com/categories/llp.htm



R. G. Brice, *Wittgenstein's On Certainty: Insight and Method*, SpringerBriefs in Philosophy, https://doi.org/10.1007/978-3-030-90781-5_7

Although unsure about using the method *himself*, Moore knew, firsthand, how effective it was when Wittgenstein used it, as he says, "he certainly has had the effect of making me very distrustful about many things which, but for him, I should have been inclined to assert positively" (Moore 1942: 33).

Distrust in what you know (or *think* you know) can be frustrating.[2] Moore's feeling underscores this. However, frustration can also serve as a catalyst for closer inspection of one's beliefs—beliefs that, although sometimes problematic and/or inaccurate, can now be adjusted and reshaped. Therapeutic philosophy requires courage in the face of such intellectually uncomfortable realizations; it also demands a certain amount of humility. One must be able to admit that a belief (or set of beliefs) one once held is (are) inaccurate or wrong. This can be challenging. As Wittgenstein once put it: "[n]othing is so difficult as *not* deceiving oneself" (Wittgenstein 1984: 34e, my emphasis).

While internal battles can be challenging, so too can external ones. Consider the challenge of going against a discipline that has, for centuries, endorsed a belief in the superiority of *one* methodological approach. Closely aligned with the sense of superiority that the Cartesian Method brings is a feeling of righteous indignation to anyone who suggests something different.[3] To reject the stringency of the Cartesian Method may be taken as a sign of laxity. Hubris runs very deep in philosophy. Philosophers who endorse the Common Sense Approach are seen as no longer taking the discipline seriously.

Take, for instance, the insults and attacks heaped upon Moore. Max Black, a contemporary philosopher of Moore's, described his philosophy as childlike: "After the intoxication of metaphysics," Black wrote, "it is good to look upon the world again as a child might-to be told 'After all, this *is* a hand. I have a body, so have you, and there are many other people like both of us who can say the same'" (Black 1950: 7). Other contemporaries of Moore, e.g., A.J. Ayer, said that Moore's "chief service to philosophy was [like] that of the child in Hans Andersen's story: he saw and was not afraid to say that the Emperor had no clothes" (Ayer 1979: 149-150). The next generation of philosophers who reflected on Moore made similar criticisms. In *The Significance of Philosophical Skepticism*, for example, Barry Stroud found what Moore said to be "perfectly acceptable" from a *non*-philosophical point of view, but "irrelevant to ... philosophical questions" (Stroud 1984: 120). Such remarks made Moore and his doctrine sound immature and adolescent. As S. Jack Odell says in his book, *On Moore*, "[m]ost philosophers are not only dismissive regarding [common sense's] validity; they are largely contemptuous of it" (Odell 2001: 48).

Interestingly, Common Sense philosophers and advocates of *Cognitive Behavioral Therapy* (CBT), share something in common. Beck describes how some of his colleagues in the mental health community tell the "troubled person" that

[2] "We just do not see how very specialized the use of 'I know' is," says Wittgenstein. "—For 'I know' seems to describe a state of affairs which guarantees what is known, guarantees it as a fact. One always forgets the expression 'I thought I knew'" (Wittgenstein 1972: §§ 11, 12).

[3] Still worse, if one argues the Cartesian Method is *wrong*.

> [h]e can't help himself and must seek out a professional healer….[B]ecause his own notions are dismissed as shallow and unsubstantial,…[this] debas[es] the value of common sense…[and] inhibits him from using his own judgment in analyzing and solving his problems (Beck 1979: 9).[4]

Beck describes the reluctance that many in his discipline have toward his (or any) new idea; a conviction that only the methodology *they* endorse is the proper way to proceed.

> [T]he mental health field is dominated by a few durable establishments and clusters of smaller sects of more tenuous standing. The major schools within this domain share certain characteristics: a conviction of the ultimate truth of their own system, disdain for opposing theories, and a steadfast emphasis on purity of doctrine and technique. In many instances, the popularity of a particular system seems to depend more on the charisma and single-mindedness of its originator than on the soundness of its foundations (Beck 1979: 7).

It remains a challenge to abandon our old ways of thinking. Our old way of thinking in both the philosophic *and* the therapeutic communities, while erudite and often highly theoretical, is, after all what is deemed "normal functioning." Reshaping these old beliefs takes time, something like this must "run its natural course" before the patient can emerge less troubled, less obsessed.[5]

The intellectual struggle *within* is demanding because it requires the patient hold on to a toxic philosophical thought while trying to reconcile it with other, new and more sensible ones. The intellectual struggle *without* is also demanding because *it* requires convincing an entire discipline/mental health field that their methodological approach is misguided.

## 7.2 Therapeutic Philosophy: Our Most Potent Inoculant

Given Wittgenstein's directive, "we may not terminate a disease of thought" (Wittgenstein 1967: §382), it is important to acknowledge that therapeutic philosophy's treatment will not always be successful. Achieving a positive and healthy philosophical mind requires a dedicated, long-term commitment.[6] Because the toxic or misleading thought must "run its natural course," because Wittgenstein prescribes

---

[4] As it turns out, this deprogramming, paradoxically, requires a trained CBT therapist to step in and re-establish the parameters of Common Sense that have been lost along the way. This is not to say that Common Sense always gets it right, either in philosophy or in therapy. When Common Sense fails to offer something plausible or useful, it is time to begin searching for what is missing. The missing data helps fill in what we do not know. After the data is located, Common Sense tools can be used for understanding and for treatment.

[5] Where he is free of an intellectual OCD—an "intellectual obsessive-compulsive disorder." My thanks to Martin Benjamin for this turn of phrase.

[6] And, in this way, philosophical therapy is different from CBT. According to the Mayo Clinic, "CBT is generally considered short-term therapy—ranging from about five to 20 session." See "Length of Therapy" https://www.mayoclinic.org/tests-procedures/cognitive-behavioral-therapy/about/pac-20384610

a "*slow* cure" (Wittgenstein 1967: §382), it would be incorrect to assume it will work for everyone. However, like CBT, for many, *therapeutic philosophy* can be an effective tool in helping individuals learn how to manage misleading and inaccurate beliefs. When exposed and filtered through our therapeutic lens, many philosophical beliefs can be reshaped into something positive and healthy. When exposed and filtered through our therapeutic lens, we stand a better chance of returning to the ancient Greek's goal of *transforming the life of the philosopher*. For Wittgenstein, this can be achieved when our "thoughts are at peace" (Wittgenstein 1984: 43e).

In *On Certainty*, Wittgenstein's philosophical and therapeutic aims dovetail nicely. His philosophical therapy aims at fostering a new way of thinking about intractable philosophical problems by encouraging philosophers to re-examine old patterns of thought and reconfigure these old patterns into new ones. Success in achieving these objectives is accomplished through a conscientious reshaping of older ideas, and patiently allowing anything toxic to "run its natural course."

We are now in a better position to address the question posed at the beginning of our investigation, namely: can Wittgenstein's therapeutic philosophy help inoculate us from the Cartesian "philosophical disease"?

It should be clear by now that therapeutic philosophy not only *can* help us, but when carefully and conscientiously applied, it is our most potent inoculant. The blind alley of Cartesianism *is* curable, but in order to treat this "philosophical disease," we must recognize when, where, and *why* it is likely to develop. When we possess these answers, we will know how best to *contain* it. Let's begin with the *when*.

### 7.2.1 When Cartesianism Develops

Early on, we discussed *Pyrrhonian* skepticism. This variety of skepticism insists that, because we cannot ascertain how things truly are in the world, we must "suspend judgment." Suspension, it was believed, would return us to a happier point in our lives, a time *before* we were "disturbed by philosophical questions" (Gutshmidt 2020: 105). Cartesian skepticism, however, marks a very different kind of disturbance, less an annoyance than a full-fledged psychiatric impairment. Descartes' methodological skepticism injected such an impossibly high level of radical doubt into our inquiry it all but destroyed our Common Sense response to it. He hoped his method would establish something "firm and lasting in the sciences," however, it quickly metastasized beyond innocent questions about whether our senses are deceiving us or whether we are dreaming, to such impractical and removed questions like, "Does the earth exist?" "Is my body real?" "Are there minds other than my own?"

Therapeutic philosophy serves as our best defense against these divorced-from-reality questions because it provides reasonable, Common Sense parameters that serve to anchor our doubts. Therapeutic philosophy's long-term goal is to prevent such questions from being taken seriously; in the short-term, however, the goal is simply to help treat those who have been infected by this "disease of thought." The patient and therapist work through the necessary steps[7] together, before the patient, at some later date, is able to use these tools on his own to ward off extraordinary skeptical questions ("extraordinary" in the sense of detached from practical or human reality.) We might think of this analogously to the way biological vaccines stimulate a person's immune system to resist a specific disease. Philosophical therapy must work to stimulate a person's Common Sense in order to identify where healthy skeptical questions end and where unhealthy Cartesian skeptical questions begin.

Albert Ellis, one of the founders of CBT, noticed that when an irrational belief is internalized, patients have a tendency to repeat the belief to themselves. After some time, the belief becomes *fixed* in their minds. Ellis likens these irrational beliefs to "negative *autosuggestions*" (Ellis 1962: 276). For this reason, as the patient and the cognitive therapist begin walking through strategies, the very first step is to "monitor…negative automatic thoughts, or cognitions" (Beck et al. 1979: 4). Monitoring, evaluating, and ultimately *modifying* negative, irrational thoughts are the steps that make up CBT. These strategies, when combined with the reasonable parameters of Common Sense, can help the patient to establish a "new normal." We are, as Ellis says:

> …uniquely suggestible as well as uniquely rational animal[s]. Other animals are to some degree suggestible and reasoning, but man's better equipped cerebral cortex, which makes possible his ability to talk to himself and others, gives him unusual opportunities to talk himself into *and* out of many difficulties (Ellis 1962: 104).

If this "talking into and out of difficulties" can be carefully calibrated with the right mix of philosophical methods and therapeutic strategies, we might successfully return some sense of reasonable balance to the patient.

### 7.2.2 Where & Why Cartesianism Develops

Much of Western philosophy has enthusiastically adopted Descartes' form of skepticism, despite Common Sense struggles against it. Indeed, Bertrand Russell, writing over 250 years *after* Descartes, was quite taken by his method:

> Descartes…invented a method which may still be used with profit—the method of systematic doubt….By inventing the method of doubt…Descartes performed a great service to

[7] Recall, in CBT the patient and therapist together work through a series of steps:

1. Identify troubling situations or conditions in your life.
2. Become aware of your thoughts, emotions and beliefs about these problems.
3. Identify negative or inaccurate thinking.
4. Reshape negative or inaccurate thinking.

These steps, as well as what one can expect at each, can be found on the Mayo Clinic's website: https://www.mayoclinic.org/tests-procedures/cognitive-behavioral-therapy/about/pac-20384610

> philosophy, and one which makes him still useful to all students of the subject (Russell 1912: 7-8).

Beck's characterization of how, in the therapeutic community, "the popularity of a particular system...depends more on the charisma and single-mindedness of its originator than on the soundness of its foundations," could, it seems to me, just as easily be directed at Descartes in philosophy.

Descartes' method has succeeded in flattering our egos by highlighting our abstract, disembodied intellect at the expense of practical action. And we, in turn, have been captivated by our own astuteness; we are mesmerized by our own sense of intellectual worth and significance. Wittgenstein, however, warned us against becoming "bewitch[ed] by our intelligence" (Wittgenstein 1953: §109). Throughout much of his writing, whether here, in *On Certainty* with his rejection of Descartes' methodological skepticism, or earlier, in *Philosophical Investigations*, with his dismantling of the very idea of a private-language, Wittgenstein has been warning us not to be taken in by Descartes. Having ignored Wittgenstein's counsel, if we wish to "get better," if we wish to improve our mental well-being, we must confront these skeptical questions in a different way, through a sort of "combative therapy," as Hamilton described it. This kind of therapy requires a strong counter-offensive, one that highlights the importance of reason, science, and Common Sense, over elevating the importance of mere logical possibility. If *this* doesn't work, however, there is one last way to address the problem of Cartesian skepticism: we work to contain the disease.

## 7.3 Quarantine: Containing the Spread of the Disease

Containing the spread of the "philosophical disease" means restricting its movement. This can be accomplished in two ways. First, following Wittgenstein's lead, we must lower epistemology from its exalted position as "one of the core areas of philosophy."[8] Rather than elevating it—at least the Cartesian version of it—as a beacon of Enlightenment thinking, we must see the theory of knowledge for what it is: just another language-game—but one among many. Therapeutic philosophy can assist us to see this more clearly.

Second, epistemology is routinely taught *through* a Cartesian lens. If we wish to cure it, this too must be contained. Descartes ought to be taught in the same way we teach Johannes Kepler...an interesting historical figure, but one who ultimately failed to use his methodology effectively. It seems reasonable to assume that as long as the Cartesian method is considered by many to be the beginning point of *all* inquiry, as long as it is believed to be the *only* epistemic method out there,

[8] *Concise Routledge Encyclopedia of Philosophy*, 2000. New York: Routledge. s.v. "epistemology." p. 246.

Wittgenstein's approach will continue to be overlooked. This, I believe, is the heart of the problem.

Convincing the academic philosopher that inquiries into what can reasonably be known as opposed to what can merely be logically conceived will be difficult, for the academic philosopher has spent his entire academic life thinking *this* is "normal functioning." Stanley Cavell put the problem this way:

> in the philosopher's context, though it may seem odd or forced, [the "reasonableness" of his grounds for doubt] does not seem *absurd*, ignorable; and because if the philosopher's request for a basis is accepted as a real question, then the bases he offers are the right, or anyway the only, bases which would seem natural (Cavell 1979: 191).

However, if philosophy is to be a worthwhile and productive discipline, if it is to be more than thought experiments and mental exercises, it must work to "change the sensibility of the patient." The goal of therapeutic philosophy must be to change the philosopher's self-defeating patterns of thinking, and this means changing his behavior. What is needed is not just his recognition that any *particular* belief is unhealthy, what is needed is a recognition that his entire *pattern of thinking* is unhealthy—a *recognition* that this thinking does not work (at least not in any practical way). If we can succeed in this goal, it might serve as a catalyst for change in behavior. This change requires a willingness to *reshape* the old way of thinking into something that is not only healthy, new, and practical, but something that will serve to *become* his "new normal."[9]

## 7.4 What We Yearn For: "Thoughts That Are At Peace"

Throughout much of this book, I have attempted to connect the strategies found in CBT to those of Wittgenstein's therapeutic method. Both methods have a similar goal: effectively transition the patient from an unhealthy condition to a healthy one. In the end, therapeutic philosophy can only be successful if the philosopher, pursuing obviously unanswerable questions, recognizes that the patterns of thinking he is familiar with are flawed and impractical. He must sincerely give up investigations into logically permissible, but at the same time *humanly impossible*,[10] lines of inquiry in favor of investigations based firmly in Common Sense. As Wittgenstein puts it in *Culture and Value*: "[t]houghts that are at peace. That's what someone who philosophizes yearns for" (Wittgenstein 1984: 43e). This sort of "peace," however, does *not* mean one stops asking questions, it *does* mean one must stop asking certain *kinds* of questions, the kind that lead one "over the edge"—the sort Wittgenstein labeled, signs of "mental disturbance."

[9] This speaks directly to CBT's goal. Again, see the Mayo Clinic's website: https://www.mayoclinic.org/tests-procedures/cognitive-behavioral-therapy/about/pac-20384610

[10] By "humanly impossible," I mean it is incompatible with actual living or leading a human life. It is "humanly impossible" if it rules out indispensable human commitments or activities.

In the end, Wittgenstein's aim is similar to that of Socrates: self-knowledge means one must engage in self-questioning and self-criticism.[11] Those who acquire this self-knowledge, those that "know thy-selves," see that their peace, their happiness, depends on psychological integration—a wholeness that sometimes requires bringing together disparate beliefs in order to reshape them into something new, something healthy—an internal peace.

We all strive to be at peace with ourselves. Inner conflict is a threat to this peace, a threat to this happiness. Given Wittgenstein's own personal obsession with certain philosophical questions, it seems he was familiar with this threat to happiness. His philosophical therapy, however, promises more than mere hope. Philosophical therapy can deliver a Common Sense cure for what plagues us in philosophy.

## References

Ayer, A.J. 1979. *Part of My Life*. London: Collins Press.

Beck, Aaron. 1979. *Cognitive Therapy and the Emotional Disorders*. New York: Penguin Books.

Beck, Aaron, John Rush, Brian Shaw, and Gary Emery. 1979. *Cognitive Therapy of Depression*. New York: Guilford.

Beck, Aaron, Gary Emery, and R. Greenberg. 2005. *Anxiety Disorders and Phobias: A Cognitive Perspective*. Cambridge: Basic Books.

Black, Max. 1950. *Philosophical Analysis: A Collection of Essays*. Ithaca: Cornell University Press.

Cavell, Stanley. 1979. *The Claim of Reason*. Oxford: Oxford University Press.

Ellis, Albert. 1962. *Reason and Emotion in Psychotherapy: A Comprehensive Method of Treating Human Disturbance*. Secaucus: Citadel.

Gutshmidt, Rico. 2020. Beyond Quietism: Transformative Experience in Pyrrhonism and Wittgenstein. *The International Journal for the Study of Skepticism* 10 (2): 105–128.

Moore, G.E. 1942. An Autobiography. In *The Philosophy of G.E. Moore*. La Salle: Open Court Publishing Company.

Odell, S. Jack. 2001. *On Moore*. Belmont: Wadsworth Publishing.

Russell, Bertrand. 1912. *The Problems of Philosophy*. Oxford: Oxford University Press.

Stroud, Barry. 1984. *The Significance of Philosophical Skepticism*. New York: Clarendon Press.

Wittgenstein, Ludwig. 1953. *Philosophical Investigations*, trans. G.E.M. Anscombe. Oxford: Blackwell Publishers.

———. 1967. *Zettel*, 2nd ed, eds. G.E.M. Anscombe and G. H. von Wright, trans. G. E. M. Anscombe. Oxford: Blackwell Publishers Ltd.

———. 1972. *On Certainty*, eds. G.E.M. Anscombe and G.H. Von Wright, trans. D. Paul and G.E.M. Anscombe. New York: Harper & Row.

———. 1984. *Culture and Value*, eds. G.E.M. Anscombe and G.H. von Wright, trans. Peter Winch. Chicago: University of Chicago Press.

[11] Recall, it is also worth noting that Aaron Beck, one of the pioneers of CBT, said his cognitive therapy "primarily used the Socratic method" (Beck et al. 2005: 167).

# Appendix

## Language-Games

Early on, in *Philosophical Investigations*, Wittgenstein compares our acquisition of language to learning the rules of a game:

> think of the whole process of using words…as one of those games by means of which children learn their native language….Think of much of the use of words in games like ring-a-ring-a-roses (Wittgenstein 1953: §7).

His example here is telling. In the game, the child may not understand the individual words in ring-a-ring-a-roses, but given the correlation between words and the embedded activities in the game—i.e., going 'round and 'round, like a "ring," falling to the ground as everyone utters "ashes, ashes, we all fall down"—the child comes to understand their meaning.[1]

The game metaphor is particularly apt. We come to understand the *meaning* of words by learning how to *use* them in particular contexts, just as we learn how to play chess by learning how the pieces can be moved in certain contexts. While identifying individual pieces in the game is, by itself, important, it does nothing to help us understand how the game is played.

> When one shows someone the king in chess and says: "This is the king," this does not tell him the use of this piece—unless he already knows the rules of the game up to this last point (Wittgenstein 1953: §31).

Only when the student has some idea of the rules regulating the movement of pieces on the board, that players try to win by capturing the opponent's pieces, can we say he understands the *meaning* of the teacher's statement, "This is the king." The

[1] Wittgenstein also draws our attention to the way in which there seems to be a broader understanding, even before we have mastered the individual concepts used in the game. Consider how a child sometimes responds by holding up three fingers when an adult asks, "How old are you?" The child does not understand that each finger represents a year—or even what a year is!



R. G. Brice, *Wittgenstein's On Certainty: Insight and Method*, SpringerBriefs in Philosophy, https://doi.org/10.1007/978-3-030-90781-5

meaning of "king" in chess is not determined by matching word and object. The meaning of "king" in chess is determined when the student knows how to correctly *use* this object in the game. A word is like a move in the game of language; the word does little for us outside of the context in which it is correctly used. "[T]he meaning of a word," Wittgenstein tells us, "is its use in the language" (Wittgenstein 1953: §43).

In the opening passages of *Philosophical Investigations*, Wittgenstein moves from simple, or as he calls them, "primitive" examples of *language-games* to more and more complex ones. Beginning in §2, he asks us to imagine a language that serves as communication for a tribe of builders.

> A is building with building-stones: there are blocks, pillars, slabs and beams. B has to pass the stones, and that in the order in which A needs them. For this purpose they use a language consisting of the words "block," "pillar," "slab," "beam." A calls them out; —B brings the stone which he has learnt to bring at such-and-such a call. —Conceive this as a complete primitive language (Wittgenstein 1953: §2).

In such an environment, Wittgenstein wonders how a child might be taught this language and how we would judge whether he has succeeded in mastering it. As part of his training process, the child must not only make *associations* between *word* and *object*, he must also learn *how* to use the word and respond correctly to other people's uses within the context of this building activity. Wittgenstein puts it this way:

> The children are brought to perform *these* actions, to use *these* words as they do so, and to react in *this* way to the words of others (Wittgenstein 1953: §6).

Learning *some* parts of language only by identifying what they refer to is not wrong, but because it neglects the way training *with the language* is embedded in this tribe's *practice* of building, it remains incomplete.[2] Functioning in its natural environment, the builder tribe's language is interwoven into the actions, the practical lives of how they *use* it. In order to see this, Wittgenstein suggests we expand the "slab," "block" language of §2 to include numerals, color samples, and words for things like "there" and "this," while used in connection with a pointing gesture. At §21 he poses the following question: "Now what is the difference between the *report or statement* "Five slabs" and the *order* "Five slabs!?" That is, what is the difference between meaning "Five slabs" as a *report* and meaning "Five slabs" as an *order*? The answer, he tells us, has to do with the *role* it is playing particular language-game:

> …it is the part which uttering these words plays in the language-game. No doubt the tone of voice and the look with which they are uttered, and much else besides, will also be different (Wittgenstein 1953: §21).

Consider its context, what *surrounds it*. What is happening before the words are uttered, what happens after? Contextual awareness helps determine how the words

[2] Wittgenstein says: "For a *large* class of cases—though not for all—in which we employ the word 'meaning' it can be defined thus: the meaning of a word is its use in the language. And the *meaning* of a name is sometimes explained by pointing to its *bearer*" (Wittgenstein 1953: §43). As the final part of this passage makes clear, it is not that meaning is *always* use, but it is far too limiting to say it is simply a one-to-one correspondence between word and object.

are to be understood. If, while constructing a house, Builder A says to Assistant B, "Five slabs," it seems reasonable to assume that A wants B to bring him five slabs. Here, it is an *order*. But if A and B were walking around the building site, say before construction begins, and A, while pointing to a pile of slabs says to B, "Five slabs," it seems likely he is giving a *report* or an inventory of what they have available. In this way, the meaning of the words is a function of the role they play *in the language-game*; it is in the patterns of activity they elicit.

Language-games are Wittgenstein's way of trying to get us away from thinking of language as a strict or "fixed" system of sentences and *towards* thinking of it as connected with *an activity*.[3]

> There are *countless* kinds: countless different kinds of use of what we call "symbols," "words," "sentences." And this multiplicity is not something fixed, given once for all; but new types of language, new language-games, as we may say, come into existence, and others become obsolete and get forgotten (Wittgenstein 1953: §23).

In §23, he offers a fuller explication of what he means by language-game and its relationship to another technical term, "form of life."

> Here the term "language-game" is meant to bring into prominence the fact that the *speaking* of language is part of an activity, or of a form of life (Wittgenstein 1953: §23).

Language is essential to, and embedded in, structured activities. These activities constitute a culture, a "form of life." Learning our language means we are participating in our culture. Wittgenstein gives some examples of language-games that constitute *our* form of life, these include:

Giving orders, and obeying them—
Describing the appearance of an object, or giving its measurements—
Constructing an object from a description (a drawing) —
Reporting an event—
Speculating about an event—
Forming and testing a hypothesis—
Presenting the results of an experiment in tables and diagrams—
Making up a story; and reading it—
Play-acting—
Singing catches—
Guessing riddles—
Making a joke; telling it—
Solving a problem in practical arithmetic—

[3] There are, of course, formal, grammatical differences between sentences, and this has led philosophers to suppose that there are basic types of sentences—i.e., assertions, questions, and commands which correspond to three distinct grammatical forms, declaratives, interrogatives, and exclamatories, respectively. Wittgenstein is not denying this, rather, his emphasis is on the *role* that the utterances play *in language-games*. He is prompting us to look at the different ways in which we actually *use* sentences. And when we look at how sentences are *used*—at the distinct language-games we play with the sentences of our language—we are then faced not with three types, but with a *countless number*.

Translating from one language into another—
Asking, thanking, cursing, greeting, praying (Wittgenstein 1953: §23).

For *less* characteristic language-games, one has to look through Wittgenstein's body of work. Located mostly in his later and "third" periods of thought, he discusses language-games of *knowledge*,[4] language-games of *metaphysics*,[5] *mathematical* language-games,[6] *logical* language-games,[7] language-games of *aesthetics*,[8] of *color*,[9] of *religion*,[10] there are even language-games of *lying*: "[l]ying is a

[4]"[T]he concept of knowing," says Wittgenstein "is coupled with that of the language-game" Wittgenstein 1972: §560.

[5]"A doubt about existence only works in a language-game" Wittgenstein 1972: §24.

[6]In *Remarks On The Foundations Of Mathematics* Wittgenstein says, "[m]athematics–I want to say–teaches you, not just the answer to a question, but a whole language-game with questions and answers." Also from *Remarks*, "A language-game in which there are axioms, proofs and proved propositions" Wittgenstein 1996: §73. And in Part VI, §20: "A further language-game is this: He gets asked 'How much is "365 X 428"?' And he may act on this question in two different ways. Either he does the multiplication, or if he has already done it before, he reads off the previous result." Part I Appendix III §7:" 'But may there not be true propositions which are written in this symbolism, but are not provable in Russell's system?'—'True propositions,' hence propositions which are true in another system, i.e. can rightly be asserted in another game." In Part V §42: "Imagine that you have taught someone a technique of multiplying. He uses it in a language-game." In *Philosophical Grammar*, Wittgenstein says, "The understanding of language, as of a game, seems like a background against which a particular sentence acquires meaning…It's much more like the understanding or mastery of a calculus, some- thing like the ability to multiply" Wittgenstein 1978: §11. In the *Blue and Brown Books* Wittgenstein says, "We introduce into our language games the endless series of numerals" Wittgenstein 1958: §21.

[7]From *Remarks Foundations Mathematics*: "Logical inference is part of a language game" Wittgenstein 1996: §30. "The introduction of a new rule of inference as a transition to a new language game" Wittgenstein 1996: §62.

[8]"The words we call expressions of aesthetic judgment play a very complicated role, but a very definite role, in what we call a culture of a period. To describe their use or to describe what you mean by a cultured taste, you have to describe a culture. What we now call a cultured taste perhaps didn't exist in the Middle Ages. An entirely different game is played in different ages" Wittgenstein 1967: §25.

[9]In *Remarks On Color* Wittgenstein says, here's "[a] language-game: Report whether a certain body is lighter or darker than another," §1. In *Remarks on the Philosophy of Psychology* Wittgenstein wrote, "Red is something specific; but we don't see that when we look at something red. Rather we see the phenomena that we limit by means of the language-game with the word 'red'" Wittgenstein 1988: §619. In the *Brown Book*, "Imagine this game: A shows B different patches of colors and asks him what they have in common" Wittgenstein 1958: Part II, 3. In *Zettel* "Think of the sentence: 'Red is not a mixed color' and of its function. For the language-game with colors is characterized by what we can do and what we cannot do" Wittgenstein 1967: §345. "Color-words are explained like this: 'That's red' e.g.–Our language game only works, of course, when a certain agreement prevails, but the concept of agreement does not enter into the language-game. If agreement were universal, we should be quite unacquainted with the concept of it" Wittgenstein 1967: §430.

[10]In *Philosophical Investigations*, Wittgenstein offers a list of some of the characteristic language-games that constitute our *form of life*, among them he lists "praying" Wittgenstein 1953: §23.

language-game that needs to be learned like any other one," he tells us.[11] But in each of these instances, Wittgenstein's point is that learning even the specialized language within such language-games depends on becoming acculturated *in* it; it means participating in a broad network of structured activities *within* that culture, *within* that form of life.

In summary then, "language games" illustrate the fact that learning a language is more than just correlating words with objects. The concept highlights the way language *works*, the way it functions within the active, practical lives of its speakers—language *in use*. It also draws our attention to the way in which language-games can be learned prior to having mastered individual concepts used in the game. Learning one's way around a particular activity (like baseball or a particular discipline like academic philosophy), involves *mastering* a particular language-game. In baseball, for example, its learning how to correctly use terms like "balls," "strikes," and "double-play," "stolen base," and "infield fly rule." In academic philosophy, its learning how to correctly use terms like "necessary and sufficient conditions," "*reductio ad absurdum*," "utilitarianism," and "Categorical Imperative."

[11] Wittgenstein 1953: §249.

# Index




R. G. Brice, *Wittgenstein's On Certainty: Insight and Method*, SpringerBriefs in Philosophy, https://doi.org/10.1007/978-3-030-90781-5

www.ingramcontent.com/pod-product-compliance
Lightning Source LLC
Chambersburg PA
CBHW051757271224
19574CB00006B/465

9 783030 907839